Restoration of Lost Samples in Digital Signals

Prentice Hall International Series in Acoustics, Speech and Signal Processing

Signals and Systems: An Introduction
L. Balmer

*Signal Processing,
Image Processing and Pattern Recognition*
S. Banks

Restoration of Lost Samples
R. Veldhuis

Prentice Hall International Series in Acoustics,
Speech and Signal Processing

Restoration of Lost Samples in Digital Signals

Raymond Veldhuis

Philips Research Laboratories
Eindhoven, The Netherlands

Prentice Hall
New York   London   Toronto   Sydney   Tokyo   Singapore

First published 1990 by
Prentice Hall International (UK) Ltd
66 Wood Lane End, Hemel Hempstead
Hertfordshire HP2 4RG
A division of
Simon & Schuster International Group

Printed and bound in Great Britain by
BPCC Wheatons Ltd, Exeter

Library of Congress Cataloging-in-Publication Data
are available from the publisher

British Library Cataloguing in Publication Data

Veldhuis, Raymond
Restoration of lost samples in digital signals.
1. Telecommunication systems. Digital signals.
Transmission
I. Title
621.3822

ISBN 0-13-775198-2

1 2 3 4 5   94 93 92 91 90

Contents

Preface

During transmission or storage, signals can be corrupted by errors. These errors can be additive noise as well as pulse-shaped distortions, the effects of which are only noticeable during short time intervals. In the latter case, if the signal is sampled data, groups of samples will be erroneous. These samples can be consecutive or scattered. Very often there are indications as to which samples are distorted. In those cases one can try to restore the errors. In this book some restoration methods for unknown samples with known positions are presented. These restoration methods are linear, that is they try to estimate the unknown samples as linear combinations of known neighbouring samples. They are also adaptive. This means that the weighting coefficients are adapted to the local (statistical) behaviour of the signal. The methods presented in this book can be applied as error concealment techniques for digital audio signals, speech signals and digital images.

As a basis for the restoration methods described in this book, a linear minimum variance estimation method is derived first. This is a general statistical method which can be used for every signal that is a realization of a stationary stochastic process. However, it is non-adaptive, since the signal spectrum, or equivalently the autocorrelation function, has to be known in advance. It is of theoretical interest, since it provides relations between the signal spectrum and the restoration error. The estimates for the unknown samples are weighted sums of the known samples. It is also possible to obtain them as the solutions of a system of linear equations.

After the discussion of the general linear minimum variance estimation method, five special cases are discussed separately. In these cases the signal spectrum can be parameterized and the systems of equations from which the estimates for the unknown samples can be solved follow directly from the signal parameters and the known samples. The resulting restoration methods are made adaptive by estimating the signal parameters from the incomplete data. In some cases iterative estimation procedures for parameters and unknown samples are developed because the parameters and the unknown samples cannot be estimated independently. The special cases discussed are sample restoration methods for autoregressive processes, speech signals, band-limited signals, multiple sinusoids and digital images. For all these cases results are presented in the form of graphs, tables and photographs.

This book has previously been published as a doctoral thesis [1]. Many people contributed in some way to the realization of this thesis. Some have contributed very explicitly by adding to the results; in this respect I think especially of the work done by Guido Janssen and the late Lorend Vries. Others, friends and immediate colleagues and especially Hans Peek, have contributed in a less explicit but also valuable way: by stimulating and inspiring me and by creating and maintaining a pleasant working atmosphere. I want also to thank them. Finally, I want to thank the management of Philips Research Laboratories for having given me the opportunity to turn a piece of work that I have enjoyed into a thesis and a book.

Eindhoven, 1990

Chapter 1

Introduction

1.1 The restoration problem

Signals generated at one place are often needed at another. The sciences dealing with the required transportation are *information theory* and *communication theory*. They describe the preprocessing to adapt a signal to the transportation medium, called the *channel*, the actual transportation, called *transmission*, and the postprocessing required to recover the original signal. A channel can be, for instance, a radio channel or an optical fibre. Transportation of signals in time, better known as *storage*, can be regarded as transmission and is also a subject of information and communication theory. In this respect, storage media, such as compact discs [2] or computer memories, can be seen as transmission channels; however, the word 'transmission' is reserved for geographical signal transportation. Examples of preprocessing are *channel coding* and *modulation*. A channel code is also called an *error correcting code*. Examples of postprocessing are *demodulation* and *channel decoding*. If a digital representation of a signal is transmitted, channel coding is often used to protect it against errors occurring during transmission. The basic idea behind channel coding is that additional bits, that make it possible to detect and sometimes correct a certain amount of transmission errors, are added to the signal [3]. Modulation is the conversion into a signal that can successfully be transmitted through the channel. A large variety of modulation methods, adapting both analogue and digital signals to various types of channels, exist. A well-known example of modulation is the following. An electrical audio signal, which contains frequencies in the range of 0–20 kHz, cannot be transmitted directly through a radio channel, but only after a translation to a higher frequency range. This can, for instance, be achieved with *AM modulation*.

1

Unfortunately, transmission channels and storage media often intro-
duce errors. For digital signals, channel coding is the tool used to correct,
or at least to detect, these errors. This book provides solutions for the
cases where errors introduced in digital signals during transmission or
storage are not corrected but only detected so that some of the received
samples are distorted. This happens, for instance, if the probability of
errors is so high that the correction capacity of the channel code is ex-
ceeded. The distorted samples are called the unknown or lost samples.
Because the channel errors are detected, their positions are known. The
solutions provided here consist in the computation of estimates for the
unknown samples from their neighbourhood. The process of estimating
and substituting new values for the unknown samples is called *(sample) restoration* or *interpolation*. The techniques can, of course, also be
applied on analogue signals that are corrupted by errors of a short duration, such as scratches on an analogue record. In that case, they must
be sampled before restoration and the positions of the pulse errors must
be detected in some way.

In this book, unknown samples are estimated as linear combinations
of known samples in their neighbourhood. As a consequence, the solutions do not work for signals consisting of pure data symbols, such as
ASCII characters, for which linear combining makes no sense. Signals
that can be restored are, for instance, signals from physical sources, such
as speech, music and images.

1.2   Restoration methods

The research on sample restoration methods, of which the results are described in this book, started in the early days of compact disc. Although
the audio signal on a compact disc is protected against errors caused by
scratches on the disc or imperfections in the disc material by an error
correcting code [4], it was thought that additional protection was desir-
able for those cases where the channel errors could not be corrected but
only detected.

The problem was stated as follows. A finite sequence of samples of
a digital audio signal, sampled at 44.1 kHz, is available. Some samples
are labelled as unknown or lost. They can be adjacent, this is called a
burst, or have scattered positions. The unknown samples are situated
approximately in the middle of the sequence. Examples of patterns of
unknown samples are shown in Figure 1.1. The problem was to find
replacements for the unknown samples such that no errors can be heard.

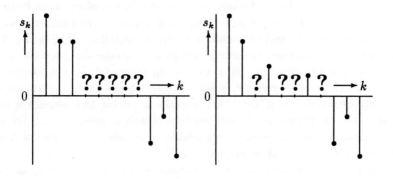

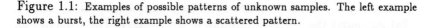

Figure 1.1: Examples of possible patterns of unknown samples. The left example shows a burst, the right example shows a scattered pattern.

Several methods have been tried. Very simple ones, such as linear interpolation or Lagrange-like curve fitting, were rejected, because they produced clearly audible errors. The first method that was more or less satisfactory was a restoration method based on the assumption that the signal was band-limited [5,6], or, in other words, that the highest frequency occurring in the signal is lower than half the sampling frequency. This method is not really successful since it gives numerical problems when the product of the number of unknown samples and the bandwidth becomes too high.[1] The best method found so far for this application is based on the assumption that the signal can be satisfactorily described with a certain signal model: the *autoregressive process* [7,8]. With this method it is possible to restore successfully bursts of up to 32 unknown samples in compact disc signals.

It so happened later that the expected problems did not occur. There are indeed uncorrectable channel errors, but the resulting number of unknown samples is usually so small that a simple restoration method such as linear interpolation suffices. Other applications for the proposed restoration method exist; it can, for instance, be used in recording studios to restore signals from scratched analogue records or from tapes with drop-outs.

When applied to speech signals, the results of this method are impressive. It is possible to restore in a speech signal, sampled at 8 kHz, bursts of up to 12.5 ms or 100 unknown samples. The method would have been a good solution to a problem occurring in mobile automatic

---

[1]This part of the research was conducted by Guido Janssen and the late Lorend Vries. The author's participation started directly after the invention of this technique.

telephony, where, due to fading, errors of this length often occur in the received signal. A problem is, however, that for this large number of unknown samples the method involves so many operations that it cannot be realized in real-time operating hardware. To evade this problem, the method was simplified by using another signal model: it was assumed that speech is *quasi-periodic*, which roughly means that it is approximately periodic, but it may vary slightly in period and waveform [9, page 184]. A definition of quasi-periodic signals is given in Chapter 4. The resulting method is extremely simple, gives good results and can be implemented in current technologies [10].

Eventually it was found that the sample restoration methods for autoregressive processes, band-limited signals and speech signals can all be regarded as special cases of a general statistical sample restoration method, that minimizes the expected restoration error energy. The estimates for the unknown samples are obtained as the solutions of a system of linear equations. The right-hand side of this system consists of linear combinations of the known samples. The weighting coefficients used to obtain the right-hand side and the coefficients of the left-hand side are directly related to the signal spectrum. For many classes of signals, including autoregressive processes, band-limited signals and quasi-periodic signals, the spectrum can be characterized or described with a few parameters. For the signal classes that have been mentioned so far, these parameters can be used to construct directly the system of equations from which the unknown samples can be solved.

Starting from this general method, it is possible to derive sample restoration methods for other signal classes, for example for sums of sinusoids [11].

In addition to the methods to restore samples in one-dimensional signals, a two-dimensional variant of the restoration method for autoregressive processes has been developed. With this method blocks of up to 8 × 8 unknown pixels in a digital image can be restored.[2] This makes the method suitable to restore errors occurring during the transmission of coded images. To reduce the bit rate, images are often coded before transmission or storage. Coding in this case is source coding [12], also called data compression or data reduction. In modern image coding systems images are often divided into blocks, for instance of 8 × 8 pixels, which are coded as a whole. The effect of an uncorrectable error is that

---

[2]A digital image is regarded as a two-dimensional array samples, called picture elements or pixels. A pixel in a monochrome image represents a local luminance value and is usually a dimensionless integer in the range 0–255. According to the CCIR 601 standard a digital television image consists of 576 lines each containing 720 pixels.

a whole block is distorted. The method proposed here can be used to restore the distorted blocks.

At the beginning of this section it was stated that the aim was to find replacements for the unknown samples in a digital audio signal such that no errors can be heard. This suggests that properties of the human observer are taken into account. This is not true; the approach in this book is signal theoretical. Models of the human auditory system exist but are hard to combine with signal theoretical results, especially because it is difficult to express the sensitivity of the human ear to restoration errors in signal theoretical terms. Human perception has been taken into account in so far as the parameter choices of the various methods have been adapted during listening tests.

Intuitively it is clear, and it will be shown later, that if the number of unknown samples increases, the restoration error also becomes larger. This means that the restoration errors will only be inaudible if the patterns of unknown samples are not too large. What is too large in this respect depends on the signal and on the restoration method.

1.3   A survey of the literature

It is surprising that the restoration problem of this book is not well covered in the signal processing literature. Most of the literature on sample restoration is on the restoration of samples in band-limited, usually low-pass, signals [13,14,5,6]. Some statistical sample restoration techniques are discussed in [15,16,17,18]. Two papers [19,20] discuss restoration methods based on waveform substitution for, respectively, speech signals and digital audio signals that are recorded on tape.

Typical examples of reconstruction methods for band-limited signals are given in [13,14,5,6]. In [13] an iterative procedure is proposed that can be used for interpolation as well as extrapolation. Initial estimates for the unknown samples are chosen, the signal is restricted to its assumed frequency band, and the signal values at the positions of the unknown samples are used as new estimates. This procedure is repeated until satisfactory results are obtained. The methods derived in [14,5,6] are essentially the same, the result is that the unknown samples are solved from a linear system of equations. In [14] this is derived starting from the method described in [13]. In [5,6] the out-of-band energy is minimized. The latter papers also give analyses of the (rather disappointing) numerical robustness of this method.

In the statistical restoration methods of [15,16,17,18] further assumptions are made about the statistics of the signal. A state space model, that has to be known in advance, is assumed in [17,18]. In [17] a restoration method based on this model is derived for continuous-time signals, and [18] discusses a restoration method for discrete-time signals of which a burst of samples is unknown. Paper [15] describes an adaptive restoration method that assumes no signal model. It deals with restoring discrete-time speech signals of which every nth sample is unknown. In [16] it is assumed that the signal is an autoregressive process with known parameters. It describes a method for restoring one unknown sample.

The waveform substitution methods of [19,20], like the restoration method for speech signals of this book, are based on the presence of a basic periodicity in the signal. Their somewhat heuristic approach is to substitute for the unknown samples a group of samples from the past, taking into account the periodicity that is present in the signal.

The methods described in [14,5,6,15,16] can all be derived from the general statistical restoration method presented in Chapter 2. The state space approach of [17,18] is somewhat different. However, as in this book, minimum variance estimates are found for the unknown samples and, provided that the same amount of available data is used, the results must be the same.

1.4   The contents of the book

In Chapter 2 a general statistical method is discussed to restore m unknown samples occurring in a signal segment of length N. As special cases, five other methods are discussed in Chapters 3–7. They are sample restoration methods for respectively: *autoregressive processes, speech signals, band-limited signals, sums of sinusoids* and *digital images*.

The method of Chapter 2 is general because no other assumptions about the available signal segment are made, other than that it is part of a realization of a stationary stochastic process [21], and that the autocorrelation function of the signal is known. The first step is the derivation of *linear minimum variance estimates* for the unknown samples. This means that the unknown samples are estimated as linear combinations of known samples, the weighting coefficients being chosen such that the estimation error has minimum variance.

After manipulating the expressions for the estimates, it is found that the weighting coefficients can be solved from a system involving the $N \times N$ autocorrelation matrix of the signal. The explicit computation of the weighting coefficients can be skipped and the m unknown samples can be solved directly from a linear system of m unknowns. This is important for the special cases discussed in this book because to construct this system and solve it requires fewer computations than to compute the weighting coefficients.

Two important cases can be distinguished. The first is that there exists one unique set of weighting coefficients, or, equivalently, a unique system of equations from which the unknown samples can be solved. This is the case if the $N \times N$ autocorrelation matrix has full rank; it is called the regular case. If the $N \times N$ autocorrelation matrix does not have full rank, it is still possible to find weighting coefficients, or a system of equations, to estimate the unknown samples. However, they may not be unique. Under certain conditions, it is then possible to make an error-free restoration. This is called the singular case. For both cases, statistical analyses of the restoration errors are given and relations with the signal spectrum are discussed. Examples of restoration methods in the regular case are those for autoregressive processes, for digital images and for speech signals. Examples of restoration methods in the singular case are those methods for band-limited signals and for sums of sinusoids.

The general restoration method described above cannot be applied very well if the $N \times N$ autocorrelation function is not known in advance. To make the method adaptive by estimating the $N \times N$ autocorrelation matrix from the data is not practical for two reasons. First, to estimate reliably the $N \times N$ autocorrelation matrix, a segment of data which is considerably larger than N samples is required [21], and the number of operations needed to compute the autocorrelation matrix would be far too high. Furthermore, the number of operations that have to be performed on the autocorrelation matrix is of the order N^2, which, in practical cases, is also too high.

In the special cases discussed in this book, the signals are modelled in some way. In this respect a signal model is not a description from which the signal can be generated completely, but a parameterization of signal properties. For instance, a signal that can be modelled as an autoregressive process can be regarded as the output of an all-pole filter, fed with white noise. The signal parameters in this case are the order of the filter, the filter coefficients and the variance of the noise. For an effective model the number of model parameters is much smaller than the amount of data. The use of models effectively reduces the number of operations required to compute estimates for the unknown samples because the system of equations from which the unknown samples are solved can be obtained directly from the model parameters and the known samples. Also, to estimate the model parameters from the available data is more efficient than to estimate the $N \times N$ autocorrelation matrix.

In the case of band-limited signals the parameters determine to which frequency bands the signal is confined. It is assumed that these are known in advance. In the other cases the parameters have to be estimated from incomplete data. In the cases of autoregressive processes, sums of sinusoids and digital images this leads to biased estimates. In those cases an iterative procedure is applied. Parameters and unknown samples are estimated in turn until a satisfactory restoration result is obtained.

The methods for restoration of unknown samples in autoregressive processes and speech signals are discussed in great detail. For these methods, analyses of numerical robustness and of the possibilities of implementations in hardware are made.

The various restoration methods are discussed in separate chapters, results, in the form of figures and tables, are presented there. The more complicated mathematical derivations are deferred to appendices.

1.5 Notations

This section gives the notational conventions used throughout the book. General notations for scalars, vectors, sets, etc. are given here. The specific symbols which are used to always denote the same parameter or quantity are given in Section 1.6.

The symbols $\mathbb{N}$, $\mathbb{Z}$, $\mathbb{R}$ and $\mathbb{C}$ denote respectively the sets of the natural numbers, the integers, the real numbers and the complex numbers. The lower and upper case roman and Greek letters are reserved for scalars. The lower case bold face roman letters are reserved for vectors and the upper case bold face roman letters are reserved for matrices. For example, a, B, γ and Δ are scalars, $\mathbf{e}$ is a vector and $\mathbf{D}$ is a matrix. The vector $\mathbf{e}$ of length n has elements

$$e_i, \quad i = 1, \ldots, n.$$

The $m \times n$ matrix $\mathbf{D}$ has elements

$$d_{i,j}, \quad i = 1, \ldots, m, \quad j = 1, \ldots, n.$$

Vector and matrix transposition is denoted by the superscript T, as in $\mathbf{e}^T$ and $\mathbf{D}^T$. The 2-norm of a vector or matrix is given by $\| \cdot \|$ [22]. The inverse of the matrix $\mathbf{M}$ is $\mathbf{M}^{-1}$. The determinant of $\mathbf{M}$ is $|\mathbf{M}|$. The inverse of the matrix $\mathbf{M}^T$ is $\mathbf{M}^{-T}$. The sum of the elements of the main diagonal of a square matrix is called the trace. The trace of the matrix $\mathbf{M}$ is denoted by $\text{trace}(\mathbf{M})$. The symbol $\mathbf{I}$ is used for the identity matrix, $\mathbf{0}$ is an all-zero vector or matrix of appropriate sizes, $\mathbf{i}_k$ is the kth unit vector, of which all elements are equal to zero, except $i_k = 1$. The null space $\mathcal{N}(\mathbf{Q})$ of an $m \times n$ matrix $\mathbf{Q}$ is defined by [22]

$$\mathcal{N}(\mathbf{Q}) = \{\mathbf{x} \in \mathbb{R}^n \mid \mathbf{Qx} = \mathbf{0}\}.$$

The range $\mathcal{R}(\mathbf{Q})$ of an $m \times n$ matrix $\mathbf{Q}$ is defined by [22]

$$\mathcal{R}(\mathbf{Q}) = \{\mathbf{y} \in \mathbb{R}^m \mid \mathbf{y} = \mathbf{Qx}, \ \mathbf{x} \in \mathbb{R}^n\}.$$

The rank of the matrix $\mathbf{Q}$ is denoted by rank($\mathbf{Q}$). The orthogonal complement of a vector space $S \subset \mathbb{R}^m$ is denoted by [22]

$$S^\perp = \left\{\mathbf{y} \in \mathbb{R}^m \mid \mathbf{y}^T\mathbf{x} = 0, \ \mathbf{x} \in S\right\}.$$

The space spanned by the vectors $\mathbf{v}_1, \ldots, \mathbf{v}_n$ is denoted by

$$\mathrm{span}\{\mathbf{v}_1, \ldots, \mathbf{v}_n\}.$$

The order of magnitude of a number q is denoted by $O(q)$. The function $\delta(x)$, $x \in \mathbb{R}$, denotes the Dirac delta function, δ_k, $k \in \mathbb{Z}$, is the Kronecker delta function.

A sampled data sequence is denoted by a lower case roman letter with a subscript. For instance s_i is the ith sample of the sampled data sequence

$$s_j, \ j = -\infty, \ldots, +\infty,$$

and

$$s_k, \ k = 1, \ldots, N,$$

is a segment taken from this sequence. A set of filter coefficients is denoted as a segment of data, for instance

$$h_l, \ l = 0, \ldots, q,$$

is possibly the set coefficients of a finite impulse response filter of order q and length $q + 1$.

Scalars, vectors, matrices and signal values can be stochastic variables, in which case they will be underlined. For instance, $\underline{\mathbf{u}}$ is a stochastic vector. Estimates for unknown samples or parameters are denoted with a circumflex. For instance, $\hat{\mathbf{a}}$ is the estimate for the parameter vector $\mathbf{a}$. The probability density functions of the stochastic variable $\underline{y}$ and of the stochastic vector $\underline{\mathbf{u}}$ are $p_{\underline{y}}(\cdot)$ and $p_{\underline{\mathbf{u}}}(\cdot)$. The statistical expectation operator is denoted by $\mathcal{E}\{\cdot\}$. The expected value, also called the mean, of stochastic variable $\underline{y}$ is denoted by μ, or μ_y, the variance of $\underline{y}$ is denoted by σ^2, or σ_y^2. The subscripts are used only if confusion would otherwise occur. In the literature different definitions exist for autocorrelation and autocovariance functions [21,23]. Here the term autocovariance function is not used. The autocorrelation function of the stationary stochastic signal $\underline{s}_i$, $i = -\infty, \ldots, +\infty$, is denoted by $R(k)$, or $R_{ss}(k)$ and defined by

$$R(k) = \mathcal{E}\left\{(\underline{s}_i - \mu)(\underline{s}_{i+k} - \mu)\right\}, \quad k = -\infty, \ldots, +\infty.$$

In most of the cases discussed in this book, μ is assumed to be equal to zero and

$$R(k) = \mathcal{E}\left\{\underline{s}_i \underline{s}_{i+k}\right\}.$$

The spectrum of the signal $\underline{s}_i$, $i = -\infty, \ldots, +\infty$, is denoted by $S(\theta)$, or $S_{ss}(\theta)$ and defined by [21]

$$S(\theta) = \sum_{k=-\infty}^{\infty} R(k) \exp(-j\theta k), \quad -\pi \leq \theta \leq \pi.$$

1.6   List of symbols

The following list summarizes the specific symbols used throughout this book. If a symbol is not found in this list, it is probably given in Section 1.5.

N	Length of the segment of available data containing the unknown samples.
s_i, $i = 1, \ldots, N$	Segment of data containing the unknown samples.
s	Vector in which the segment of data is arranged.
m	Number of unknown samples.
$t(i)$, $i = 1, \ldots, m$	Positions of the unknown samples.
v_i, $i = 1, \ldots, N$	Segment of data, with zeros substituted at the positions of the unknown samples.
v	Vector in which the segment of data is arranged, with zeros substituted at the positions of the unknown samples.

x	Vector in which the unknown samples are arranged.
$R(k)$, $k = -\infty, \ldots, +\infty$	Signal's autocorrelation function.
$S(\theta)$, $|\theta| \leq \pi$	Signal spectrum.
R	Signal's $N \times N$ autocorrelation matrix.
e	Restoration error vector.
E	Restoration error covariance matrix.
M, N	Vertical (M) and horizontal (N) sizes of a block of image data containing the unknown pixels.
$s_{i,j}$, $i = 1, \ldots, M$; $j = 1, \ldots, N$	Block of image data, containing the unknown pixels.
$(t_1(i), t_2(i))$, $i = 1, \ldots, m$	Vertical and horizontal positions of the unknown pixels.
V	Set of the indices of the unknown samples. For instance, in the one-dimensional case $V = \{t(1), \ldots, t(m)\}$.
W	Set of the indices of the available samples, including the unknown ones. For instance in the one-dimensional finite case $W = \{1, \ldots, N\}$.
$W \setminus V$	Set of the indices of the available samples, excluding the unknown ones.

Chapter 2

Linear minimum variance estimation

2.1 Introduction

In this chapter a general statistical sample restoration method for one-dimensional sampled data is presented.[1] The method is called general because no further assumptions about the signal are made than that it is *a realization of a stochastic process that is stationary up to order 2* [21] and that its autocorrelation function is known. Stationarity roughly means that the statistical properties of the stochastic signal are independent of time. A stochastic signal

$$\underline{s}_i, \ i = -\infty, \ldots, +\infty,$$

is stationary up to order 2 if the first and second order moments

$$\mathcal{E}\left\{\underline{s}_j\right\}, \ \mathcal{E}\left\{\underline{s}_j\underline{s}_{j+k}\right\}$$

are independent of j. Precise definitions of (complete) stationarity and stationarity up to order p are given in [21]. Stationarity is a more severe requirement than stationarity up to a certain finite order, but many stochastic signals are not even stationary up to order 2. Fortunately, in this book it is sufficient to assume that the signal is stationary up to order 2 on a segment of length N. This type of stationarity is called *local stationarity*. For N not too high, this requirement can be met in many practical cases. The maximum value of N for which a segment can be considered stationary depends on the kind of stochastic signal.

---

[1]Parts of the contents of this chapter have been published previously in [11].

The segment of data that is available is denoted by

$$s_j, \quad j = 1, \dots, N.$$

This segment contains m unknown samples, at the known positions

$$t(i), \quad i = 1, \dots, m.$$

In this chapter no further restrictions are put on the positions of the unknown samples.

In Section 2.2 linear minimum variance estimates are derived for the unknown samples. The term linear refers to the fact that the unknown samples are estimated as linear combinations of the known ones. The estimates are called minimum variance estimates because they are chosen such that the variance of the estimation error is minimized. It is shown that the weighting coefficients can be solved from Wiener-Hopf-like equations [23]. These are rearranged in a more convenient form, giving more insight into the relation of the coefficients to the signal's autocorrelation function. It is shown that their explicit computation can be avoided and that the unknown samples can be directly obtained as the solutions of a system of m linear equations with m unknowns. The right-hand side of this system consists of linear combinations of the known samples. The weighting coefficients used to obtain the right-hand side of the system and the coefficients of the left-hand side are directly related to the signal spectrum. For the special cases that are discussed in Chapters 3–7 the signal spectra can be described or characterized with only a few parameters. In those cases these parameters can be used directly to construct the system of equations. This approach is computationally more efficient than the computation of one set of weighting coefficients for every unknown sample.

In Section 2.3 interesting relations with the signal spectrum are derived under the assumption that the segment of available data has infinite length. This section is of more than just theoretical importance, because its results are sometimes also valid if just a finite amount of data is available. Section 2.4 discusses the restoration error and the restoration error covariance matrix. Section 2.5 summarizes the main results of this chapter and gives conclusions.

2.2 Derivation of estimators

Let s_i, $i = 1, \dots, N$ be the available segment of data with unknown samples at positions $t(1), \dots, t(m)$. Let V, W and $W \setminus V$ denote respectively the set of indices of the unknown samples, the set of the indices of all the available samples including the unknown ones and the set of the known samples. The available segment is part of a realization of a stochastic process $\underline{s}_i$ that has zero mean. Coefficients

$$h_{i,j}, \ i = 1, \ldots, m, \ j \in W \setminus V,$$

have to be computed so that $s_{t(1)}, \ldots, s_{t(m)}$ are estimated by

$$\hat{s}_{t(i)} = \sum_{j \in W \setminus V} h_{i,j} s_j, \ i = 1, \ldots, m. \tag{2.1}$$

There is just one set of coefficients $h_{i,j}$ and there are infinitely many realizations of $\underline{s}_i$. In order to get the average best performance, the $h_{i,j}$ must be optimized over all the realizations. This can be done by choosing the coefficients such that they minimize the total variance of the statistical restoration error, defined by

$$\mathcal{E} \left\{ \sum_{i=1}^{m} (\hat{\underline{s}}_{t(i)} - \underline{s}_{t(i)})^2 \right\}. \tag{2.2}$$

Note that in (2.2) estimates and original data are stochastic variables. The $\hat{\underline{s}}_{t(i)}$ follow from (2.1) by replacing s_j by $\underline{s}_j$. The stochastic estimates $\hat{\underline{s}}_{t(i)}$ are called *estimators*. On substitution of (2.1) into (2.2) one obtains for the total variance of the statistical restoration error the expression

$$\sum_{i=1}^{m} \left(\sum_{j,k \in W \setminus V} h_{i,j} h_{i,k} R(j - k) - 2 \sum_{j \in W \setminus V} h_{i,j} R(j - t(i)) + \sigma_s^2 \right). \tag{2.3}$$

Here $R(\cdot)$ is the autocorrelation function of the stochastic signal $\underline{s}_i$ and σ_s^2 is the signal variance. Expression (2.3) is quadratic in the $h_{i,j}$, and if there is a minimum, it can be found by setting the derivatives with respect to the $h_{i,j}$ equal to zero. The $h_{i,j}$ that minimize (2.3) then follow as solutions of m resulting sets of linear equations, each with $N - m$ unknowns, given by

$$\sum_{j \in W \setminus V} h_{i,j} R(k - j) - R(t(i) - k) = 0, \ k \in W \setminus V, \ i = 1, \ldots, m. \tag{2.4}$$

The index i numbers the sets of equations, the index k numbers the equations. These equations show a great resemblance to the *Wiener-Hopf equation* [23]. The general approach in estimation problems of using linear estimates to minimize error variance has been popular ever since its introduction by Wiener around 1940 [24] and, at the same time but independently, by Kolmogorov [25]. The reason for its popularity is probably that the weighting coefficients can be obtained easily as the solutions of a set of linear equations.

If the coefficients $h_{i,j}$ can be solved from (2.4), this chapter on linear minimum variance estimation can stop at this point. However, this still has to be shown. Furthermore, manipulating these equations a little leads to results that provide more insight. Note that the $h_{i,j}$ can be regarded as the coefficients of an $m \times N$ matrix $\mathbf{H}$ of which not all elements have been specified. To define this matrix $\mathbf{H}$ completely, choose for the unspecified coefficients

$$h_{i,j} = \left\{ \begin{array}{ll} -1, & \text{if } i = 1, \ldots, m, \ j = t(i), \\ 0, & \text{if } i = 1, \ldots, m, \ j \in V, \ j \neq t(i). \end{array} \right. \tag{2.5}$$

The $N \times N$ autocorrelation matrix $\mathbf{R}$ of the stochastic signal $\underline{s}_i$ is defined by

$$r_{i,j} = R(i - j), \ i, j = 1, \ldots, N. \tag{2.6}$$

Autocorrelation matrices are special matrices: they are *symmetric, nonnegative definite* and *Toeplitz* [22]. These are all matrix properties which will be of use later. Some of the properties of autocorrelation matrices are discussed in Appendix A. The $(N - m) \times N$ matrix $\mathbf{R}'$ is a submatrix of $\mathbf{R}$, obtained by deleting the rows with indices $t(1), \ldots, t(m)$. The equations (2.4) can now be written down in a matrix form

$$\mathbf{R}'\mathbf{H}^T = 0, \tag{2.7}$$

where 0 is the $(N - m) \times m$ all-zero matrix.

It must be verified whether an $m \times N$ matrix $\mathbf{H}$ satisfying (2.5) and (2.7) always exists. For least squares problems, where expressions of the form

$$\| \mathbf{Ax} - \mathbf{b} \|^2$$

are minimized, it can be shown that there is always a solution, though possibly not a unique one [22,26]. Here, the expression (2.3) that is minimized is of the form

$$\mathbf{x}^T \mathbf{Cx} + 2\mathbf{c}^T \mathbf{x} + d,$$

and for arbitrary non-negative definite $\mathbf{C}$ and $\mathbf{c}$ it is not clear in advance whether there is a solution.[2] In Appendix B, however, it is shown that in this case there always exists a, possibly not unique, matrix $\mathbf{H}$. In this

---

[2]This problem has a solution if and only if $\mathbf{c} \in \mathcal{R}(\mathbf{C})$.

appendix some other properties of the solutions of the system (2.7) are also given. It must be remarked that the fact that a set of weighting coefficients can always be found does not guarantee good restoration results. This will become clear from Example 1.

The m rows of $\mathbf{H}$ belong to the *null space* $\mathcal{N}(\mathbf{R}')$ of $\mathbf{R}'$ [22]. If $\mathbf{R}'$ has full rank, then $\mathcal{N}(\mathbf{R}')$ has dimension m, otherwise it has a dimension greater than m. The proof given in Appendix B that a matrix $\mathbf{H}$ satisfying (2.5) and (2.7) exists is based on the fact that $\mathcal{N}(\mathbf{R}')$ always contains a set of m independent vectors $\mathbf{g}_1, \ldots, \mathbf{g}_m$ that can be arranged in an $m \times N$ matrix

$$\mathbf{G} = \begin{bmatrix} \mathbf{g}_1^T \\ \vdots \\ \mathbf{g}_m^T \end{bmatrix} \tag{2.8}$$

with a full rank $m \times m$ submatrix $\tilde{\mathbf{G}}$, defined by

$$\tilde{g}_{i,j} = g_{i,t(j)}, \quad i = 1, \ldots, m, \quad j = 1, \ldots, m. \tag{2.9}$$

As a consequence, $\mathbf{H}$ follows from

$$\mathbf{H} = -\tilde{\mathbf{G}}^{-1}\mathbf{G}. \tag{2.10}$$

Example 1 *Consider the stochastic signal*

$$\underline{s}_k = \underline{a} \cos(k\frac{\pi}{2} + \underline{\phi}), \quad k = -\infty, \ldots, +\infty.$$

The amplitude $\underline{a}$ and the phase $\underline{\phi}$ are stochastic, $\mathcal{E}\{\underline{a}^2\} = \sigma_a^2$. The phase has a uniform probability distribution on the interval $[-\pi, \pi)$. Assume that a segment of length 3 is available, of which the second sample is unknown. The matrices $\mathbf{R}$ and $\mathbf{R}'$ are given by

$$\mathbf{R} = \frac{\sigma_a^2}{2} \begin{bmatrix} 1 & 0 & -1 \\ 0 & 1 & 0 \\ -1 & 0 & 1 \end{bmatrix}, \quad \mathbf{R}' = \frac{\sigma_a^2}{2} \begin{bmatrix} 1 & 0 & -1 \\ -1 & 0 & 1 \end{bmatrix}.$$

The vectors $[0, 1, 0]^T$ and $[1, 0, 1]^T$ are a basis for the null space of $\mathbf{R}'$, and

$$\mathbf{H} = [a, -1, a],$$

where a is an arbitrary scalar. It is easily seen that the estimate for the unknown second sample in any sequence of 3 samples is always 0. From (2.38) in Section 2.4 it follows that the expected restoration error is $\sigma_a^2/2$, which equals the signal variance. This illustrates that even though a set of weighting coefficients can be found, the interpolation results are poor.

Example 2 *Consider the stochastic signal*

$$\underline{s}_k = \underline{a}\cos(k\frac{\pi}{6} + \underline{\phi}), \quad k = -\infty, \ldots, +\infty.$$

The amplitude $\underline{a}$ and the phase $\underline{\phi}$ are as in Example 1. Assume that a segment of length 3 is available, of which the second sample is unknown. The matrices $\mathbf{R}$ and $\mathbf{R}'$ are given by

$$\mathbf{R} = \frac{\sigma_a^2}{2}\begin{bmatrix} 1 & \frac{1}{2}\sqrt{3} & \frac{1}{2} \\ \frac{1}{2}\sqrt{3} & 1 & \frac{1}{2}\sqrt{3} \\ \frac{1}{2} & \frac{1}{2}\sqrt{3} & 1 \end{bmatrix}, \quad \mathbf{R}' = \frac{\sigma_a^2}{2}\begin{bmatrix} 1 & \frac{1}{2}\sqrt{3} & \frac{1}{2} \\ \frac{1}{2} & \frac{1}{2}\sqrt{3} & 1 \end{bmatrix}.$$

The vector $[-\frac{1}{\sqrt{3}}, 1, -\frac{1}{\sqrt{3}}]^T$ is a basis for the null space of $\mathbf{R}'$, and

$$\mathbf{H} = \left[\frac{1}{\sqrt{3}}, -1, \frac{1}{\sqrt{3}}\right].$$

It is easily seen that the estimate for the unknown second sample in any sequence of 3 samples is always correct: the expected restoration error is 0. The reason will become clear in Section 2.4.

The matrix $\mathbf{G}$ plays an important role in this book. It will be shown later in this section and in Section 2.3 how the elements of $\mathbf{G}$ can be computed directly from the signal's $N \times N$ autocorrelation matrix or, equivalently, from the signal spectrum. In the special cases of the general sample restoration method of this chapter, that are discussed in the following chapters, the signal spectrum is described or parameterized with only a few parameters. The approach is, in all those cases, to derive the matrix $\mathbf{G}$ directly from these parameters.

The calculation of the matrix $\mathbf{H}$ that contains the weighting coefficients in (2.10) can be skipped, and the unknown samples can be solved directly from a system of equations. If $\mathbf{G}$ is computed directly from the signal's parameters, this approach saves computations, because solving a system of equations requires fewer operations than inverting a matrix, as is required in (2.10). The segment of available samples s_i, $i = 1, \ldots, N$, can be arranged in a vector $\mathbf{s}$. Let $\mathbf{v}$ be the vector obtained from $\mathbf{s}$ by substituting zeros in the positions of the unknown samples, and finally, let $\hat{\mathbf{x}}$ be a vector of estimates for the unknown samples defined by

$$\hat{x}_i = \hat{s}_{t(i)}, \quad i = 1, \ldots, m. \tag{2.11}$$

Then $\hat{\mathbf{x}}$ follows from

$$\hat{\mathbf{x}} = \mathbf{H}\mathbf{v} = -\tilde{\mathbf{G}}^{-1}\mathbf{G}\mathbf{v}. \tag{2.12}$$

Multiplying both sides of (2.12) by $\tilde{\mathbf{G}}$ gives

$$\tilde{\mathbf{G}}\hat{\mathbf{x}} = -\mathbf{G}\mathbf{v} \stackrel{\text{def}}{=} -\mathbf{z}, \qquad (2.13)$$

from which the unknown samples can be solved directly. The m vector $\mathbf{z}$ is called the *syndrome*, its elements are linear combinations of the known samples. This approach is especially convenient in the adaptive restoration methods of Chapters 3–7. Then the coefficients of the system (2.13) are obtained directly from signal parameter estimates, and the number of operations required to compute $\hat{\mathbf{x}}$ directly from (2.13) is less than via the intermediate computation of $\mathbf{H}$.

It will now be investigated how $\mathbf{G}$ is related to the signal's auto-correlation function.

The rows of $\mathbf{G}$ (2.8) belong to $\mathcal{N}(\mathbf{R}')$. Since $\mathcal{N}(\mathbf{R}) \subset \mathcal{N}(\mathbf{R}')$, they belong either to $\mathcal{N}(\mathbf{R})$ or to a subset T of $\mathbb{R}^N$, defined by

$$T = \qquad\qquad\qquad\qquad\qquad\qquad\qquad\qquad (2.14)$$
$$\left\{ \mathbf{x} \mid (\mathbf{R}\mathbf{x})_i = 0,\ i \in W \setminus V \wedge [(\mathbf{R}\mathbf{x})_{t(1)}, \ldots, (\mathbf{R}\mathbf{x})_{t(m)}]^T \neq \mathbf{0} \right\}.$$

Two interesting cases can be distinguished. The first is called the *regular case*. If the $N \times N$ autocorrelation matrix $\mathbf{R}$ has full rank, $\mathbf{G}$ can be derived from

$$\mathbf{R}\mathbf{G}^T = [\mathbf{i}_{t(1)}, \ldots, \mathbf{i}_{t(m)}]. \qquad (2.15)$$

The rows of $\mathbf{G}$ are columns of the inverse of the $N \times N$ autocorrelation matrix and $\tilde{\mathbf{G}}$ has elements

$$\tilde{g}_{i,j} = (\mathbf{R}^{-1})_{t(i),t(j)},\ i,j = 1, \ldots, m. \qquad (2.16)$$

Since it was assumed that $\mathbf{R}$ has full rank, $\mathbf{R}$ is positive definite and has a positive definite inverse. From (2.16) it follows that $\tilde{\mathbf{G}}$ is a principal submatrix of $\mathbf{R}^{-1}$ and is also positive definite [27]. In this case the rows of $\mathbf{G}$ all belong to T. Note that in Example 1 the rows of $\mathbf{G}$ also belong to T, but that $\mathbf{R}$ does not have full rank.

The second case of interest is when $\mathbf{R}$ has a rank less than or equal to $N - m$. Then, if the rows of $\mathbf{G}$ can all be chosen from $\mathcal{N}(\mathbf{R})$, $\mathbf{G}$ must be a non-trivial solution of

$$\mathbf{R}\mathbf{G}^T = \mathbf{0}. \qquad (2.17)$$

This is called the *singular case*. For both the regular and the singular cases elegant expressions for the restoration error are derived in Section 2.4. Note that in Examples 1 and 2 the $\mathbf{R}$s both have rank 2, but in Example 1 the row of **G** cannot be chosen from $\mathcal{N}(\mathbf{R})$, whereas in Example 2 it can.

It is shown in Appendix A that an autocorrelation matrix of finite dimensions is only singular if it is the autocorrelation matrix of a stochastic signal consisting of a finite number r of sinusoids. In this case the autocorrelation matrix has maximum rank $2r$, independent of its dimensions. In all other cases an autocorrelation matrix of finite dimensions is regular. Note that in Examples 1 and 2 the signals consist of one sinusoid and the rank of the autocorrelation matrices is two. In practice, however, it can happen that for increasing N the $N \times N$ autocorrelation matrix becomes almost singular. As is also shown in Appendix A, this happens if the signal spectrum $S(\theta)$ is zero over some subinterval of $[0, \pi]$ or, in other words, if the signal is in some sense band-limited. This behaviour can also be explained with the *Szegö limit theorem* [28,29], from which it can be concluded that for $N \to \infty$ the eigenvalue distribution of **R** and the value distribution of $S(\theta)$, $-\pi \le \theta \le \pi$, coincide. Therefore, for increasing N, **R** will become almost singular. The Szegö limit theorem is used again and clarified in Section 2.4.

2.3 Infinite segments of data

The number of samples used to estimate the unknown samples has so far been finite. If **R**, **G** and **H** are allowed to be infinite matrices, the previously obtained results apply also for the estimation of unknown samples in an infinite sequence. In that case some relations with the signal spectrum can be derived. Even though it is unrealistic to assume that an infinite sequence of data is available, the results of this section have practical value. If **R** is infinite, usually **G** will also be infinite. In the cases discussed in this book, however, even for an infinite **R** a finite **G** can be found, or **G** can be approximated by a finite matrix. In this section only the regular and the singular cases, as defined in Section 2.2, are considered.

In the regular case the jth row $\mathbf{g}_j$ of **G** follows from

$$\mathbf{R}\mathbf{g}_j = \mathbf{i}_{t(j)}.$$

If **R** is infinite, this can be written as a convolution

$$\sum_{k=-\infty}^{\infty} R(k)(\mathbf{g}_j)_{l-k} = \delta_{l-t(j)}. \tag{2.18}$$

In the singular case the jth row $\mathbf{g}_j$ of **G** follows from

$$\mathbf{R}\mathbf{g}_j = \mathbf{0}.$$

If **R** is infinite, this can be written as a convolution

$$\sum_{k=-\infty}^{\infty} R(k)(\mathbf{g}_j)_{l-k} = 0. \qquad (2.19)$$

In both the regular and the singular cases the rows of **G** become shifted versions of a sequence g_k, and the elements of **G** satisfy

$$g_{i,j} = g_{j-t(i)}. \qquad (2.20)$$

In the regular case g_k, $k = -\infty, \ldots, +\infty$, follows from

$$g_k = \frac{1}{2\pi} \int_{-\pi}^{\pi} \frac{1}{S(\theta)} \exp(j\theta k) \, d\theta, \quad k = -\infty, \ldots, +\infty. \qquad (2.21)$$

Or, equivalently,

$$
\begin{aligned}
G(\exp(j\theta)) &= \sum_{k=-\infty}^{\infty} g_k \exp(-j\theta k) \\
&= \frac{1}{S(\theta)}.
\end{aligned}
\qquad (2.22)
$$

In the singular case g_k, $k = -\infty, \ldots, +\infty$, follows from

$$\frac{1}{2\pi} \int_{-\pi}^{\pi} G(\exp(j\theta)) \, S(\theta) \, \exp(j\theta k) \, d\theta = 0. \qquad (2.23)$$

The submatrix $\tilde{\mathbf{G}}$ of **G** follows from

$$\tilde{g}_{i,j} = g_{t(j)-t(i)}, \quad i, j = 1, \ldots, m. \qquad (2.24)$$

In the case of bursts of unknown samples, $\tilde{\mathbf{G}}$ is Toeplitz. The syndrome **z**, defined in (2.13) is the result of a convolution

$$z_i = \sum_{k=-\infty}^{\infty} g_{k-t(i)} \, v_k, \quad i = 1, \ldots, m. \qquad (2.25)$$

Here v_k, $k = -\infty, \ldots, \infty$, is the available signal with zeros substituted for the unknown samples.

Usually the sequence g_k of (2.21) or (2.23) has infinite length, but if $S(\theta)^{-1}$ is a finite length polynomial in $\exp(j\theta)$, as is the case for autoregressive processes of finite order and for quasi-periodic speech signals, then g_k is also of finite length. In that case, if on either side of the pattern of unknown samples there are enough correct samples, the results of this section apply. If the length of g_k is $2p + 1$, then on either side of the unknown samples there must be at least p correct samples. In Appendix A it is shown that if the signal consists of a finite number of sinusoids, g_k is also of finite length, and if the signal is band-limited, g_k can be approximated with the impulse response of a finite length selective filter.

The sequence g_k of (2.21) has some interesting properties that will help in understanding the adaptive restoration methods of Chapters 3, 4 and 7. If $S(\theta)$ is a rational function in $\exp(j\theta)$, its z-transform $S(z)$ can be written as[3]

$$S(z) = \sigma^2 \frac{U(z)}{V(z)} \frac{U(z^{-1})}{V(z^{-1})}. \tag{2.26}$$

Here $U(z)$ and $V(z)$ are polynomials in z with leading coefficients equal to one, that have their zeros inside the unit circle of the z-plane. Consequently $U(z^{-1})$ and $V(z^{-1})$ have their zeros outside the unit circle of the z-plane, σ^2 is a positive constant. As a consequence of this, g_k can be written as

$$g_k = \frac{1}{\sigma^2} \sum_{l=0}^{\infty} a_l a_{l+k}, \quad k = -\infty, \ldots, +\infty, \tag{2.27}$$

where a_l is a sequence, defined by

$$a_l = \begin{cases} 0, & l < 0, \\ 1, & l = 0, \\ \frac{1}{2\pi j} \oint_{|z|=1} \frac{V(z)}{U(z)} z^{l-1} dz, & l > 0. \end{cases} \tag{2.28}$$

The sequence a_k, $k = 0, \ldots, \infty$, can be obtained by minimizing as a function of a_k, $k = 1, \ldots, \infty$, the expression

$$\mathcal{E} \left\{ \left(\sum_{k=0}^{\infty} a_k \underline{s}_{l-k} \right)^2 \right\}. \tag{2.29}$$

The result is a system of equations, in the finite case known as the *Yule-Walker* equations [30,31],

$$\sum_{k=1}^{\infty} a_k R(l - k) = -R(l), \quad l = 1, \ldots, \infty. \tag{2.30}$$

In the adaptive cases discussed in Chapters 3, 4 and 7 estimates for (2.29), obtained from a finite segment of data, are minimized as a function of a finite number of coefficients a_k. The results are sets of equations similar

---

[3]The notations $S(z)$ and $S(\theta)$ are inconsistent, but will be used only in this section.

to (2.30), involving estimates for the $R(k)$. The g_k are then computed by using (2.27). Now consider the expression

$$\sum_{l=-\infty}^{\infty} \left(\sum_{k=0}^{\infty} a_k s_{l-k} \right)^2 . \tag{2.31}$$

This expression can be written in the form

$$\sum_{l=-\infty}^{\infty} \left(\sum_{k=0}^{\infty} a_k v_{l-k} \right)^2 + 2\sigma^2 \mathbf{x}^T \mathbf{z} + \sigma^2 \mathbf{x}^T \tilde{\mathbf{G}} \mathbf{x}. \tag{2.32}$$

Only the second and the third term of (2.32) involve unknown samples; $\mathbf{z}$ is defined in (2.25). The first, infinite, term involves only known samples. It is clear that the solution $\hat{\mathbf{x}}$ of (2.13) also 'minimizes' (2.32). In the adaptive cases discussed in Chapters 3, 4 and 7 a similar expression, obtained from a finite segment of data, is minimized as a function of $\hat{\mathbf{x}}$. In those cases the estimate for (2.29) and the finite version of (2.32) are given by the same expression.

2.4   The restoration error

The linear minimum variance estimates derived in the previous section were chosen such that their overall performance is statistically optimal over the realizations of $\underline{s}_i$, $i = -\infty, \ldots, +\infty$. Since the actual restoration error depends on the realization, a general statement on the restoration error must be in statistical terms. In this section expressions are presented for the variance of the restoration error, defined in (2.2), and expressions for the *restoration error covariance matrix* $\mathbf{E}$ are derived. If $\underline{\mathbf{x}}$ and $\hat{\underline{\mathbf{x}}}$ are vectors containing the samples at positions $t(1), \ldots, t(m)$ and their estimators, then the statistical restoration error $\underline{e}$ is given by

$$\underline{e} = \hat{\underline{\mathbf{x}}} - \underline{\mathbf{x}}, \tag{2.33}$$

and $\mathbf{E}$ is defined by

$$\mathbf{E} = \mathcal{E}\left\{ (\hat{\underline{\mathbf{x}}} - \underline{\mathbf{x}})(\hat{\underline{\mathbf{x}}} - \underline{\mathbf{x}})^T) \right\} = \mathcal{E}\left\{ \underline{ee}^T \right\}. \tag{2.34}$$

The variance of the restoration error is given by

$$\mathcal{E}\left\{ \sum_{i=1}^{m} (\hat{\underline{s}}_{t(i)} - \underline{s}_{t(i)})^2 \right\} = \text{trace}(\mathbf{E}) = \underline{e}^T \underline{e}. \tag{2.35}$$

The estimator $\hat{\underline{x}}$ is given by $\hat{\underline{x}} = \mathbf{H}\underline{v}$ (2.12). By using (2.5) this can be written as

$$\hat{\underline{x}} = \mathbf{H}\underline{s} + \underline{x}. \tag{2.36}$$

It follows from

$$\mathcal{E}\{\underline{e}\} = \mathbf{H}\mathcal{E}\{\mathbf{s}\} = 0 \tag{2.37}$$

that the estimators are unbiased. On substitution of (2.36) into (2.34) and by using (2.10) one obtains

$$
\begin{aligned}
\mathbf{E} &= \mathcal{E}\left\{\mathbf{H}\underline{s}\underline{s}^T\mathbf{H}^T\right\} \\
&= \mathbf{H}\mathbf{R}\mathbf{H}^T \\
&= \tilde{\mathbf{G}}^{-1}\mathbf{G}\mathbf{R}\mathbf{G}^T\tilde{\mathbf{G}}^{-T}. \tag{2.38}
\end{aligned}
$$

In the regular case, on substitution of (2.15), it follows that

$$\mathbf{E} = \tilde{\mathbf{G}}^{-1}. \tag{2.39}$$

In the singular case, on substitution of (2.17), it follows that

$$\mathbf{E} = 0. \tag{2.40}$$

In the singular case an error-free restoration is possible. In the regular case the restoration error depends via (2.16) on the signal's autocorrelation function. It must be remarked at this point that, even though theoretically an error-free restoration in the singular case is possible, the restoration results can be disappointing. The quality of the restoration is determined not only by the statistical restoration error variance, but also by the numerical robustness of the applied restoration method. In the adaptive cases of Chapters 3–7, model parameters are estimated from the available data, and with these parameters the system (2.16) is constructed. Errors will occur in the estimated parameters and in the constructed system. Their influences on the restoration results depend greatly on the sensitivity to errors of $\hat{x}$ in (2.16), which is determined by the *condition number* of $\tilde{\mathbf{G}}$ [22,32]. It turns out, for example, that the 'error-free' restoration method for band-limited signals of Chapter 5 is much more sensitive to errors than the restoration method for autoregressive processes of Chapter 3 that is not 'error-free'.

Intuitively it can be understood that if m increases, the restoration error in the regular case increases as well. It is also intuitively clear that the error variance of each restored sample will always be less than or equal to the signal's variance $R(0)$ This is because an error variance $R(0)$ is already obtained by substituting zeros for the unknown samples, and linear minimum variance estimates will, by nature, at least have the same performance. For a burst of unknown samples and a Toeplitz $\tilde{G}$, which happens if the segment of data is infinite or if g_k from (2.21) has finite length and the unknown samples are embedded in a sufficiently large amount of known samples, some results on the variance of restoration error can be derived.[4]

The first result that can be obtained is a symmetry property of the error variance of the restored samples. If $\tilde{G}$ is Toeplitz, it is also persymmetric, which means that

$$g_{i,j} = g_{m+1-j,m+1-i}.$$

It is a property of persymmetric matrices that their inverses are also persymmetric. Therefore, the restoration error covariance matrix $\mathbf{E}$ is persymmetric, and in particular $e_{i,i} = e_{m+1-i,m+1-i}$, or

$$\mathcal{E}\left\{\underline{e}_i^2\right\} = \mathcal{E}\left\{\underline{e}_{m+1-i}^2\right\}. \tag{2.41}$$

This shows the symmetrical behaviour of the error variances of the restored samples in the case of bursts of unknown samples.

If the eigenvalues of $\tilde{G}$ are denoted by $\lambda_1, \ldots, \lambda_m$, with

$$\lambda_1 \leq \ldots \leq \lambda_m,$$

then the total restoration error variance is given by

$$\text{trace}(\mathbf{E}) = \text{trace}(\tilde{G}^{-1}) = \sum_{i=1}^{m} \frac{1}{\lambda_i}. \tag{2.42}$$

According to the Szegö limit theorem [28,29], for any function $F(\cdot)$, continuous on the set

$$\left\{G(\theta) = \sum_{k=-\infty}^{\infty} g_k \exp(-j\theta k), \ |\theta| < \pi\right\},$$

one has

$$\lim_{m\to\infty} \frac{1}{m} \sum_{i=1}^{m} F(\lambda_i) = \frac{1}{2\pi} \int_{-\pi}^{\pi} F(G(\theta)) \, d\theta. \tag{2.43}$$

Taking $F(a) = a^{-1}$, and by using (2.22), one finds

$$\lim_{m\to\infty} \frac{1}{m} \sum_{i=1}^{m} F(\lambda_i) = \frac{1}{2\pi} \int_{-\pi}^{\pi} S(\theta) \, d\theta = R(0). \tag{2.44}$$

---

[4]These results are also derived in [7] for the special case of autoregressive processes.

Hence,

$$\lim_{m\to\infty} \frac{1}{m} \mathcal{E}\left\{\underline{e}^T \underline{e}\right\} = R(0). \tag{2.45}$$

This shows that for long bursts of unknown samples the error variance per sample approaches the signal variance, which is also expected intuitively.

The restoration error can be analyzed in more detail if the stochastic signal $\underline{s}_k$, $k = -\infty, \ldots, +\infty$, is Gaussian; then $\underline{e}$ has a probability density function

$$p_{\underline{e}}(e) = \frac{|\tilde{\mathbf{G}}|^{\frac{1}{2}}}{(2\pi)^{\frac{m}{2}}} \exp\left(-\frac{e^T \tilde{\mathbf{G}} e}{2}\right). \tag{2.46}$$

It is a rather tedious but straightforward exercise to find for the variance of $\underline{e}^T \underline{e}$ the expression

$$\mathrm{var}(\underline{e}^T \underline{e}) = 2 \sum_{i=1}^{m} \frac{1}{\lambda_i^2}. \tag{2.47}$$

For large m, by using the Szegö limit theorem (2.43) with $F(a) = a^{-2}$ one finds

$$\lim_{m\to\infty} \frac{1}{m} \mathrm{var}(\underline{e}^T \underline{e}) = 2\, \frac{1}{2\pi} \int_{-\pi}^{\pi} (S(\theta))^2 \; d\theta. \tag{2.48}$$

2.5 Conclusions

In this chapter it has been shown that linear minimum variance estimators for the values of unknown samples in a segment of sampled data can always be derived if all of the following hold:

- The positions of the unknown samples are known.

- The segment is taken from a realization of a stochastic process that is stationary up to order 2 on this segment.[5]

- The $N \times N$ autocorrelation matrix of the signal, where N is the segment length, is known.

---

[5]This is not a necessary condition. Linear minimum variance estimates can still be obtained if the signal is not stationary. In that case the autocorrelation coefficients $R(k - j)$ and $R(t(i) - k)$ in (2.3) and (2.4) must be replaced by the autocovariance coefficients $C(k, j) = \mathcal{E}\{\underline{s}_k \underline{s}_j\}$ and $C(t(i), k) = \mathcal{E}\{\underline{s}_{t(i)} \underline{s}_k\}$, respectively. The results of Section 2.2 and Appendix B remain mostly valid if the autocorrelation matrix $\mathbf{R}$ is replaced by the autocovariance matrix $\mathbf{C}$, with $c_{i,j} = \mathcal{E}\{\underline{s}_i \underline{s}_j\}$, $i, j = 1, \ldots, N$. For non-stationary signals the relations with the signal spectrum can no longer be made.

The estimates for the unknown samples are linear combinations of the known samples. In this chapter it has been shown how the weighting coefficients that are required to make these linear combinations can be obtained. Two methods have been presented. The first is to derive the weighting coefficients directly from the $N \times N$ autocorrelation matrix. The second is to build a system of equations from which the unknown samples can be solved. The left-hand side of this system depends on the autocorrelation matrix, the right-hand side depends linearly on the known samples and also on the autocorrelation matrix. The latter case is the basis for the restoration methods discussed in Chapters 3–7.

Two interesting cases have been discussed. In the first case, called the regular case, the $N \times N$ autocorrelation matrix has full rank. In the second case, called the singular case, the $N \times N$ autocorrelation matrix has a rank less than or equal to $N - m$, where m is the number of unknown samples. In the latter case the restoration error is always equal to zero. The case where the $N \times N$ autocorrelation matrix has a rank between $N - m$ and N has not been discussed.

The case of the available segment of data having infinite length has been discussed. This seems an unpractical case, but it will turn out for the cases discussed in Chapters 3–7 that, even though the segment is assumed to be of infinite length, only a finite number of known samples are required to estimate the unknown samples. Furthermore, in this case interesting relations with the signal spectrum can be derived.

The general restoration method, as it has been presented in this chapter, is not practical because it requires knowledge of the signal's $N \times N$ autocorrelation matrix. Interesting signals, such as audio and video signals, have statistics that vary in time. In those cases the autocorrelation matrix is unknown and has to be estimated. Unfortunately, for a reliable estimate of the $N \times N$ autocorrelation matrix a segment is required with a length that is much greater than N. Estimation of the autocorrelation matrix in this case requires many operations. A second problem is that to obtain the weighting coefficients or to build the system of equations from which the unknown samples can be solved, the autocorrelation matrix has to be inverted. This also requires many operations. For these reasons the method presented in this chapter must not be seen as a practical one; it is merely a theoretical tool that can be used for analysis purposes. Nevertheless it provides the basis for the other restoration methods presented in Chapters 3–7. One of the results of this chapter was that the unknown samples can always be obtained as the solutions of a set of equations. In the cases discussed in Chapters 3–7 the signal spectrum can be described or characterized with only a few parameters that can be estimated directly from the available data. It will be seen in Chapters 3-7 that the system of equations can be derived directly from these parameters and the known samples. The adaptive restoration procedure that is followed in those cases is first to estimate the parameters,[6] second to build the system of equations, and third to solve this system. In the cases discussed in Chapters 3 and 6 the parameters cannot be estimated reliably because the segment contains unknown samples. This problem is overcome by making the restoration method iterative.

---

[6]The method of Chapter 5 on sample restoration in band-limited signals is an exception. There it is assumed that the bandwidth is known.

Chapter 3

Autoregressive processes

3.1 Introduction

In this chapter an adaptive restoration method for unknown samples in autoregressive processes is derived.[1] The autoregressive process is important in signal modelling. Many real-life signals can be modelled successfully as autoregressive processes. For instance in the fields of spectral estimation [33,30] and speech coding [34] the autoregressive process is often used as a signal model. The results of this chapter, presented in Section 3.8, show that for sample restoration purposes it is also a good model for music signals.

The organization of this chapter is as follows. Section 3.2 gives the definition of an autoregressive process and gives expressions for its spectrum. In Section 3.3 linear minimum variance estimators for the unknown samples are derived. This comes down, in fact, to finding the sequence g_k from (2.21) for an autoregressive process. For autoregressive processes some remarks on the restoration error, additional to the ones of Section 2.4, can be made. This is done in Section 3.4. In adaptive cases the parameters of the autoregressive process are generally unknown. Therefore, in Section 3.5 the method is made adaptive. The result is an iterative procedure, in which alternately the parameters and the unknown samples are estimated from the available data. This method is related to another iterative statistical estimation procedure, the EM algorithm [35,36,37] and to Newton-Raphson's method. The relations are discussed in Appendix E. Section 3.6 discusses computational aspects. The autoregressive model is a special case of the more general *autoregressive-moving average* model. The restoration of autoregressive-moving average processes is not discussed in this book. The reason in given in Section 3.7.

---

[1]The contents of this chapter have been published previously in [7].

28

Results are presented in Section 3.8.

3.2 Autoregressive processes

It is assumed that s_k, $k = -\infty, \ldots, +\infty$, is a realization of a stationary autoregressive process $\underline{s}_k$, $k = -\infty, \ldots, +\infty$. This means that there exist a finite positive integer p, called the *order of prediction*, real numbers $a_0, a_1, \ldots, a_p$, $a_0 = 1$, called the *prediction coefficients*, and a zero-mean white noise process $\underline{e}_k$, $k = -\infty, \ldots, +\infty$, called the *excitation noise*, with variance σ_e^2, such that

$$\sum_{l=0}^{p} a_l \underline{s}_{k-l} = \underline{e}_k, \quad k = -\infty, \ldots, +\infty. \tag{3.1}$$

No further assumptions are made about the probability density function of the $\underline{e}_k$, $k = -\infty, \ldots, +\infty$, than that they are continuous zero mean stochastic variables. For notational convenience, it shall be agreed that $a_k = 0$ for $k < 0$ or $k > p$. The spectrum $S(\theta)$ of the autoregressive process of (3.1) is given by

$$
\begin{aligned}
S(\theta) &= \frac{\sigma_e^2}{\left|\sum_{l=0}^{p} a_l \exp(-\mathrm{j}\theta l)\right|^2} \\
&= \frac{\sigma_e^2}{\sum_{l=-p}^{p} b_l \exp(-\mathrm{j}\theta l)},
\end{aligned}
\tag{3.2}
$$

where

$$b_l = \sum_{k=0}^{p} a_k a_{k+l}. \tag{3.3}$$

An autoregressive process can be regarded as the output of an all-pole filter with a transfer function

$$\frac{1}{A(\exp(\mathrm{j}\theta))} = \frac{1}{\sum_{l=0}^{p} a_l \exp(-\mathrm{j}\theta l)},$$

excited with zero-mean white noise with variance σ_e^2. An example with a third order filter is shown in Figure 3.1. It can be seen that this filter generates samples $\underline{s}_k$ satisfying (3.1) if this expression is rewritten as

$$\underline{s}_k = \underline{e}_k - \sum_{l=1}^{p} a_l \underline{s}_{k-l}, \quad k = -\infty, \ldots, +\infty. \tag{3.4}$$

Expression (3.4) and Figure 3.1 also illustrate another way to look at autoregressive processes. The right term of the right-hand side of (3.4) and

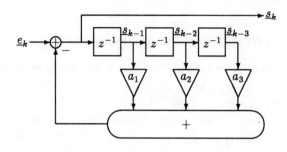

Figure 3.1: Example of a third order autoregressive process as the output of an all-pole filter excited by white noise.

the output of the large adder in Figure 3.1, both taken with a minussign, can be regarded as a prediction of $\underline{s}_k$, based on the p previous samples. In that case the excitation noise sample $\underline{e}_k$ is the prediction error. This clarifies why p and the $a_0, \ldots, a_p$ are called the order of prediction and the prediction coefficients, respectively. The zeros of the polynomial

$$A(z) = \sum_{l=0}^{p} a_l z^{-l}, \ z \in \mathbb{C}, \tag{3.5}$$

are the poles of the filter. They must be within the unit circle of the complex plane.

3.3   Linear minimum variance estimates

Throughout the whole chapter, it is assumed that

$$p + 1 \le t(1) < \ldots < t(m) \le N - p. \tag{3.6}$$

This means that it is assumed that there are guard spaces of at least p correct samples on either side of the pattern containing the unknown samples. Estimates for the unknown samples for arbitrary $t(1), \ldots, t(m)$ can also be found, but the results are less elegant and also less good. On substituting (3.2) into (2.21) it follows for g_k that

$$g_k = \begin{cases} \sigma_e^{-2} b_k, & -p \le k \le p, \\ 0, & \text{otherwise.} \end{cases}$$

Because σ_e^2 appears on both sides of the system (2.13), is it just as convenient to define

$$g_k = \begin{cases} b_k, & -p \leq k \leq p, \\ 0, & \text{otherwise.} \end{cases} \tag{3.7}$$

The vector $\hat{x}$ of unknown samples can be solved from the system (2.13)

$$\tilde{G}\hat{x} = -z,$$

with

$$\tilde{g}_{i,j} = b_{t(j)-t(i)}, \quad i, j = 1, \ldots, m, \tag{3.8}$$

and

$$z_i = \sum_{k=-p}^{p} b_k v_{k-t(i)}, \quad i = 1, \ldots, m. \tag{3.9}$$

Note that here a restoration method that assumes the availability of an infinite segment of data is applied, but that the resulting sequence g_k has finite length. It is permissible to apply this method if (3.6) holds, which means that it should be guaranteed that the unknown samples are embedded in a sufficiently large neighbourhood of known ones. In general, if more known samples are used to estimate the unknown samples, statistically the restoration error will decrease. However, since the length of g_k from (2.21) is determined by the order of prediction p, if (3.6) holds, increasing N will not help any further.

3.4   The restoration error

The error covariance matrix E is given by

$$E = \sigma_e^2 \tilde{G}^{-1}, \tag{3.10}$$

with $\tilde{G}$ defined in (3.8). The relative restoration error variance per sample E, defined by

$$E = \frac{\sigma_e^2 \, \text{trace}\left(\tilde{G}^{-1}\right)}{mR(0)} \tag{3.11}$$

is a measure for the performance of the restoration method.

The following example demonstrates that the performance of this method is higher for autoregressive processes with a peaky spectrum, of which the zeros of (3.5) are close to the unit circle, than for autoregressive processes with a smooth spectrum. In Section 3.6 simulation results are presented that also illustrate this property.

Example 3 *Consider a second order autoregressive process, that has poles at positions*

$$\rho \exp(\pm j\phi), \ 0 \le \rho < 1, \ 0 \le \phi \le \pi.$$

The prediction coefficients are given by

$$a_0 = 1, \ a_1 = -2\rho \cos(\phi), \ a_2 = \rho^2.$$

There is one unknown sample at position $t(1)$. The restoration error variance is given by

$$\frac{\sigma_e^2}{b_0} = \frac{\sigma_e^2}{1 + 4\rho^2 \cos^2(\phi) + \rho^4}.$$

It is more interesting to evaluate the relative error variance per sample, which for this example is given by

$$E = \frac{\sigma_e^2}{b_0 R(0)}.$$

The signal variance $R(0)$ can be obtained by evaluating

$$R(0) = \frac{1}{2\pi j} \oint_{|z|=1} \frac{\sigma_e^2}{z A(z) A(z^{-1})} \, dz.$$

The result is

$$R(0) = \sigma_e^2 \frac{1 + \rho^2}{1 - \rho^2} \frac{1}{1 - 2\rho^2 \cos(2\phi) + \rho^4}.$$

Combining the results, one obtains

$$\frac{\sigma_e^2}{b_0 R(0)} = \frac{1 - \rho^2}{1 + \rho^2} \frac{1 - 2\rho^2 \cos(2\phi) + \rho^4}{1 + 2\rho^2 \cos(2\phi) + \rho^4 + 2\rho^2} \le \frac{1 - \rho^4}{1 + \rho^4}.$$

From this result it can be seen that if ρ approaches 1, the relative restoration error variance becomes very small. In the limit case of a signal consisting of sinusoids, the poles are on the unit circle. It is shown in Chapter 2 that in that case the restoration error is 0.

The phenomenon that autoregressive processes with a peaky spectrum can be restored with a smaller error than autoregressive processes with a smooth spectrum can also be observed for other patterns of unknown samples and higher order processes. However, a general proof for other patterns of unknown samples seems difficult. In Appendix C it is shown that for one unknown sample and an autoregressive process of order p the relative restoration error variance has the following upper bound:

$$\frac{\sigma_e^2}{b_0 R(0)} \leq \frac{1 - (\prod_{i=1}^{p} \rho_i)^2}{1 + (\prod_{i=1}^{p} \rho_i)^2}. \tag{3.12}$$

Here ρ_i is the modulus of the ith zero of $A(z)$ in the z-plane. If the ρ_i, $i = 1, \ldots, p$, are close to 1, in which case the spectrum is certainly peaky, the upper bound for the relative error variance becomes small. This result is intuitively pleasing because if a signal spectrum is flat the signal has the character of white noise and becomes unpredictable; if the spectrum is peaky some frequencies in the signal will be dominant and the signal becomes better predictable.

It is shown in Appendix C that (3.12) holds with the equality sign if and only if $A(z) = 1 - a_p z^{-p}$. In that case the autoregressive process has poles

$$\alpha_k = |a_p|^{\frac{1}{p}} \exp(\mathrm{j}\frac{2\pi}{p}k), \quad k = 1, \ldots, p,$$

if $a_p > 0$, or

$$\alpha_k = |a_p|^{\frac{1}{p}} \exp(\mathrm{j}(\frac{2\pi}{p}k + \frac{\pi}{p})), \quad k = 1, \ldots, p,$$

if $a_p < 0$. The spectrum of this process is the flattest that can be realized with an autoregressive process of order p and therefore the signal is the least predictable.

3.5   Adaptive restoration

Until now the parameters, prediction coefficients and order of prediction, were assumed to be known. In practice cases this is often not so, and then both parameters and unknown samples have to be estimated from the available, incomplete, data. A problem arises here. Methods exist for estimating the parameters from complete data [33,30], and, as has been shown in Section 3.3, there are also methods for estimating unknown samples if the parameters are known, but there are no methods known for estimating both parameters and unknown samples from incomplete data. In this section a method of doing this is developed. It is related to a statistical estimation procedure, the EM algorithm. This relation is discussed in Appendix E.

The first problem is to estimate the order p of the autoregressive process. Even in the case of complete data, order estimation is troublesome.

Some methods exist, for instance [38], but they do not work well for music signals, which was the first application of the restoration method of this chapter. Some a priori knowledge of the order of prediction must be used, but if this is not available, for music the rather heuristic choice $p \cong 3m$ gives good results for m up to 16; for higher m, p can be fixed to 50. This relation has been obtained experimentally. Its use can be motivated by the reasoning that on either side of the pattern of unknown samples, p samples are linearly combined into estimates and it seems sensible to make the amount of data proportional to the number of unknown samples.

For convenience the notation

$$\mathbf{a} = [a_1, \ldots, a_p]^T$$

is adopted. The estimation of the prediction coefficients and of the unknown samples $\mathbf{x}$ is expressed as a minimization problem, where estimates $\hat{\mathbf{a}}$ and $\hat{\mathbf{x}}$ are chosen such that

$$Q(\mathbf{a}, \mathbf{x}) = \sum_{k=p+1}^{N} \left(\sum_{l=0}^{p} a_l s_{k-l} \right)^2 = \sum_{k=p+1}^{N} e_k^2 \qquad (3.13)$$

is minimal as a function of $\mathbf{a}$ and $\mathbf{x}$. Except for a scaling factor, the expression (3.13) is an estimate for (2.29). It is also a finite approximation of (2.31). Once $\hat{\mathbf{a}}$ and $\hat{\mathbf{x}}$ are obtained, σ_e^2 is estimated by

$$\hat{\sigma}_e^2 = \frac{1}{N - p - m} Q(\hat{\mathbf{a}}, \hat{\mathbf{x}}). \qquad (3.14)$$

The particular choice for minimizing $Q(\mathbf{a}, \mathbf{x})$ to obtain estimates for $\mathbf{a}$ and $\mathbf{x}$ is further motivated by the following facts. First, assuming that the data are complete, if $\mathbf{s} = [s_1, \ldots, s_N]^T$ and $\mathbf{u} = [s_1, \ldots, s_p]^T$ then, under the hypothesis that the sample values have a Gaussian probability density function, minimizing $Q(\mathbf{a}, \mathbf{x})$ with respect to $\mathbf{a}$ turns out to be the same as maximizing the log likelihood function

$$L(\mathbf{a}, \sigma_e^2) = \log(p_{\underline{\mathbf{s}}|\underline{\mathbf{u}}}(\mathbf{s}|\mathbf{u}, \mathbf{a}, \sigma_e^2)) \qquad (3.15)$$

as a function of $\mathbf{a}$ and σ_e^2. This is a common procedure for estimating $\mathbf{a}$ and σ_e^2, and its validity is proved in Appendix D. Second, assuming that the prediction coefficients are known, as is the case with (2.31), $Q(\mathbf{a}, \mathbf{x})$ is minimized by the linear minimum variance estimates for the unknown samples, which is easily seen, since

$$Q(\mathbf{a}, \mathbf{x}) = \sum_{k=p+1}^{N} \left(\sum_{l=0}^{p} a_l v_{k-l} \right)^2 + 2\mathbf{x}^T \mathbf{z} + \mathbf{x}^T \tilde{\mathbf{G}} \mathbf{x}, \qquad (3.16)$$

with $\mathbf{z}$ from (3.9), $\tilde{\mathbf{G}}$ from (3.8). In Appendix D it is also shown that minimizing $Q(\mathbf{a}, \mathbf{x})$ as a function of $\mathbf{x}$ is equivalent to finding minimum variance estimates or maximum a posteriori estimates for the unknown samples under the assumption that the samples have a Gaussian probability density function.

Since $Q(\mathbf{a}, \mathbf{x})$ involves fourth order terms, such as $a_i^2 s_{t(m)}^2$, the minimization with respect to $\mathbf{a}$ and $\mathbf{x}$ is a non-trivial problem. Good results are obtained with the following iterative minimization procedure. One chooses an initial estimate $\hat{\mathbf{x}}^{(0)}$, for instance $\hat{\mathbf{x}}^{(0)} = \mathbf{0}$, for the unknown samples. Next, one minimizes $Q(\mathbf{a}, \hat{\mathbf{x}}^{(0)})$ as a function of $\mathbf{a}$ to obtain a first estimate $\hat{\mathbf{a}}^{(1)}$ for the vector of prediction coefficients. Then one minimizes $Q(\hat{\mathbf{a}}^{(1)}, \mathbf{x})$ as a function of $\mathbf{x}$ to obtain a first estimate $\hat{\mathbf{x}}^{(1)}$ for the vector of unknown samples. The procedure can be repeated by computing $\hat{\mathbf{a}}^{(2)}$, $\hat{\mathbf{x}}^{(2)}$, $\hat{\mathbf{a}}^{(3)}$, $\hat{\mathbf{x}}^{(3)}$ and so on, until a satisfactory restoration result has been obtained. If the algorithm is applied to digital audio, with for instance $N = 512$, $m = 16$ and $p = 50$, just the computation of $\hat{\mathbf{x}}^{(1)}$ suffices. It is clear that in this way $Q(\mathbf{a}, \mathbf{x})$ decreases to some non-negative number. One may hope that the sequence of values of $Q(\mathbf{a}, \mathbf{x})$ thus obtained converges to its global minimum. Unfortunately, it seems very hard to prove any definite result in this direction. In Example 4 a convergence result for the case $m = 1$, $p = 1$ is given. Further remarks on the convergence of this sequence are made in Appendix E. This iterative minimization procedure closely resembles a maximum likelihood parameter estimation algorithm, well-known in statistics: the EM algorithm [35,36,37]. It is also closely related to Newton-Raphson's method for minimizing functions [39]. These resemblances are discussed in Appendix E.

The two minimization steps of the iterative procedure, one with respect to the parameters, the other with respect to the unknown samples, are both feasible since $Q(\mathbf{a}, \mathbf{x})$ is quadratic in $\mathbf{a}$ for known $\mathbf{x}$ and quadratic in $\mathbf{x}$ for known $\mathbf{a}$. In fact, it can be shown that

$$Q(\mathbf{a}, \mathbf{x}) = c_{0,0} + 2\mathbf{a}^T \mathbf{c} + \mathbf{a}^T \mathbf{C}\mathbf{a}. \qquad (3.17)$$

Here

$$c_{i,j} = \sum_{k=p+1}^{N} s_{k-i} s_{k-j}. \qquad (3.18)$$

The p vector $\mathbf{c}$ from (3.17) is defined by $\mathbf{c} = [c_{0,1}, \ldots, c_{0,p}]^T$ and the $p \times p$ matrix $\mathbf{C}$ has elements $c_{i,j}$, $i, j = 1, \ldots, p$. The matrix $\mathbf{C}$ is sometimes called the *autocovariance matrix* and this method of finding prediction coefficients is known as the *autocovariance method* [30]. Minimizing (3.17)

as a function of the prediction coefficients leads to the following system
of equations:

$$\mathbf{C\hat{a}} = -\mathbf{c}, \qquad (3.19)$$

from which $\hat{a}$ can be solved with efficient methods [30]. The system
(3.19) is a finite version of (2.30), with the $R(i-j)$ replaced by estimates
$c_{i,j}$. If it is assumed that the s_k are equal to zero outside the interval
$k = 1, \ldots, N$, and the summation in (3.18) is taken over $k \in \mathbb{Z}$, then it
can be shown that

$$\frac{1}{N}c_{i,j} = \hat{R}(i-j),$$

where $\hat{R}(j)$ is a (biased) estimate for $R(j)$. The resulting estimation
method for the prediction coefficients is then called the *autocorrelation
method*, and the system (3.19) is solved by the Levinson-Durbin algorithm
[33,30,31]. A comparison of these methods is made in [33,30]. As is seen
from (3.16), $Q(\mathbf{a},\mathbf{x})$ is minimized by the solution $\hat{\mathbf{x}}$ of the system (2.13)
$\mathbf{\tilde{G}\hat{x}} = -\mathbf{z}$. On substitution of (3.19) into (3.14), it follows easily that

$$\hat{\sigma}_e^2 = \frac{1}{N-p-m}(c_{0,0} + \hat{\mathbf{a}}^T\mathbf{c}). \qquad (3.20)$$

Also, on substitution of $\underline{s}_k$ and $\underline{\hat{s}}_{t(i)} = \underline{s}_{t(i)} + \underline{e}_i$, $i = 1, \ldots, m$, into (3.14)
and by writing the $Q(\hat{\mathbf{a}}, \underline{\hat{\mathbf{x}}})$ thus obtained in the form (3.16), and by using
the expression (3.10), it can be shown that

$$\mathcal{E}\left\{\frac{1}{N-p-m}Q(\mathbf{a},\underline{\hat{\mathbf{x}}})\right\} = \sigma_e^2. \qquad (3.21)$$

Or, in words, the estimate (3.14) for σ_e^2 is unbiased.

The vector $\hat{\mathbf{x}}$ that minimizes (3.16) can also be obtained as the solution to a problem in the frequency domain. Assume that the samples s_k
equal zero outside the interval $1 \le k \le N$. It is then easily verified that
the expression

$$\sum_{k=-\infty}^{\infty}\left(\sum_{l=0}^{p}a_l s_{k-l}\right)^2 \qquad (3.22)$$

is still of the form (3.16) and that in fact only the first term has changed
slightly. Therefore, (3.22) is also minimized by the vector $\hat{\mathbf{x}}$ that minimizes (3.16). By using Parseval's equality it can be shown that

$$\sum_{k=-\infty}^{\infty}\left(\sum_{l=0}^{p}a_l s_{k-l}\right)^2 \qquad (3.23)$$

$$= \frac{1}{2\pi}\int_{-\pi}^{\pi}\left|\sum_{k=0}^{N-1}s_{k+1}\exp(-j\theta k)\right|^2 |A(\exp(j\theta))|^2 \, d\theta$$

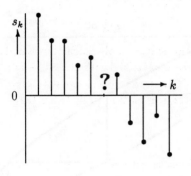

Figure 3.2: Example of a sequence.

$$= \frac{1}{2\pi} \int_{-\pi}^{\pi} \sigma_e^2 \frac{\left| \sum_{k=0}^{N-1} s_{k+1} \exp(-j\theta k) \right|^2}{S(\theta)} d\theta.$$

Intuitively, by minimizing (3.23) as a function of the unknown samples, one forces the restored signal to have little (or much) spectral energy in those regions of the frequency domain, where $S(\theta)$ is small (or large). This brings out a relation with the sample restoration method for band-limited signals of [5], discussed in Chapter 5, where the restoration is such that the spectral energy of the restored signal is concentrated as much as possible in the assumed baseband of the original signal. It should be noted that the integral in (3.23) can be related to the work of [40] on distortion measures for spectral densities.

Example 4 *This example illustrates the convergence behaviour of the restoration method of this chapter. Consider the signal segment consisting of $N = 11$ samples,*

$$s_{11} = [294, 201, 199, 110, 139, ?, 76, -101, -172, -75, -218]^T,$$

which is depicted in Figure 3.2. In this segment the sample at position $t = 6$ is unknown. The samples are taken from a realization of an autoregressive process, with $p = 1$, $a_1 = -0.9$ and $\sigma_e^2 = 1.0E + 04$. The original value of s_6 is 37. For this simple case, it is easily verified that $\hat{a}_1$, according to (3.19), follows from

$$\hat{a}_1 = -\frac{c_{0,1}^{(0)} + \hat{x}_t(x_{t-1} + x_{t+1})}{c_{1,1}^{(0)} + \hat{x}_t^2}. \tag{3.24}$$

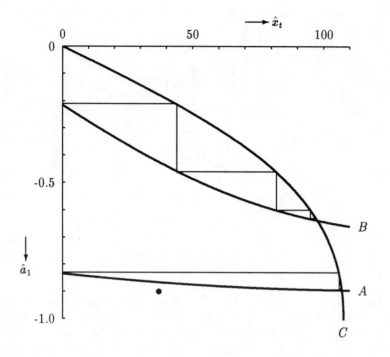

Figure 3.3: Convergence for long and short data sequences.

Here $c_{0,1}^{(0)}$ and $c_{1,1}^{(0)}$ are as $c_{0,1}$ and $c_{1,1}$ in (3.18), but with $s_t = 0$. Also, $\hat{x}_t$, according to (2.13), follows from

$$\hat{x}_t = -\frac{\hat{a}_1\left(x_{t-1} + x_{t+1}\right)}{1 + \hat{a}_1^2}. \tag{3.25}$$

The minimum of $Q(a_1, x_t)$ is at the intersection of (3.24) and (3.25). It can be checked that the curves (3.24) and (3.25) in general only intersect at one point of the $\hat{a}_1, \hat{x}_t$ plane, so that the function $Q(a_1, x_t)$ has one global minimum. Figure 3.3 shows the curves (3.24), denoted by A and B, and the curve (3.25), denoted by C for the data sequence $\mathbf{s}_{11}$, curve A, and for the sequence $\mathbf{s}_5$, curve B. The sequence $\mathbf{s}_5$ is the subsequence of length $N = 5$ of $\mathbf{s}_{11}$, given by

$$\mathbf{s}_5 = [110, 139, ?, 76, -101]^T.$$

For $\mathbf{s}_{11}$ one has $c_{0,1}^{(0)} = 175, 219$, $c_{1,1}^{(0)} = 210, 133$, and for $\mathbf{s}_5$ one has $c_{0,1}^{(0)} = 7,614$, $c_{1,1}^{(0)} = 35,298$. The curve (3.25) does not depend on N, but, via $c_{0,1}^{(0)}$ and $c_{1,1}^{(0)}$, the curve (3.24) does.

i	$\mathbf{s}_5$		$\mathbf{s}_{11}$	
	$\hat{a}_1^{(i)}$	$\hat{x}_t^{(i)}$	$\hat{a}_1^{(i)}$	$\hat{x}_t^{(i)}$
1	-0.22	44	-0.83	106
2	-0.46	82	-0.89	107
3	-0.60	95	-0.89	107
4	-0.63	97	-0.89	107
5	-0.64	97	-0.89	107

Table 3.1: Intermediate estimation results for a long and a short data sequence.

The sequences $\hat{a}_1^{(i)}$, $\hat{x}_t^{(i)}$ that are obtained when $Q(a_1, x_t)$ is minimized iteratively are shown in Table 3.1 for the sequences $\mathbf{s}_5$ and $\mathbf{s}_{11}$. The iteration steps are shown in Figure 3.3, as well as the values of (a_1, x_t) and $(a_1^{(i)}, x_t^{(i)})$.

Convergence is much faster for $\mathbf{s}_{11}$. The reason for this is that the bias in $c_{0,1}^{(0)}$ and $c_{1,1}^{(0)}$ is less because of the increased number of known samples, and consequently the initial estimate of a_1 is closer to a_1.

The final value of $\hat{x}_t$ is not very good, but starting with the true value of a_1 would have given an estimate $\hat{x}_t = 107$, identical with the value obtained with $\mathbf{s}_{11}$. As in Example 3, it is interesting to evaluate the value of the relative error variance E, defined in (3.11), for known a_1. In this case one has

$$E = \frac{\sigma_e^2}{b_0 R(0)} = \frac{1 - a_1^2}{1 + a_1^2}.$$

For $a_1 = 0.9$ the relative error variance is approximately 0.10. This indicates that for values of a_1 that are not very close to 1, as is the case here, the estimates for the unknown samples are not very reliable.

3.6 Computational aspects

In this section the computational aspects of the calculation of $\hat{\mathbf{a}}$ in (3.19) and $\hat{\mathbf{x}}$ in (2.13) are considered. This is of interest if the algorithm described in this chapter is implemented as an integrated circuit. Based on the results described in this section and on fixed-point simulations done by the author of this book, proposals for an integrated circuit have been made by two students of the University of Louvain [41].[2] They implemented a version of the algorithm capable of restoring maximally 16

---

[2]The source text of the FORTRAN program performing the fixed point simulations is included in this reference.

unknown samples in a sequence of 512 samples, assuming that the order of prediction p equals 50. They arrived at chip sizes varying between 100 mm^2 and 150 mm^2 in the 3μ n-well CMOS process.

It should be noted that linear systems need to be solved for the estimation both of prediction coefficients and unknown samples. If the order of prediction p is chosen to be approximately $3m$, then the need for efficiency is more urgent for estimating the prediction coefficients than the unknown samples.

The computation of $\hat{a}$ in (3.19) is a well-known problem. In Section 3.5, two types of approach were mentioned, the autocorrelation and the autocovariance methods. In the IC proposal [41] the autocorrelation method is used, implemented in 16-bit block-floating [42] arithmetic. For both methods efficient system-solving algorithms exist [33,30]. The number of operations required is $O(p^2)$ for both methods. Apart from these, there are other methods that can be used to compute these prediction coefficients from the data [30]. The numerical properties of some of these methods are discussed in [43]. A substantial proportion of the computations required to compute the prediction coefficients is needed to compute $\mathbf{C}$ and $\mathbf{c}$ if the autocovariance method is used, or to compute the autocorrelation function estimate $\hat{R}(j)$ if the autocorrelation method is used. In both cases the number of operations required is $O(Np)$, which is generally much more than the number of operations required to solve the system (2.13) for the prediction coefficients, but which can be reduced by using fast, number-theoretic, correlation methods. In the IC proposal of [41] a different approach has been taken. To compute a reliable autocorrelation estimate, the 16-bit precision of the signal s_k is not required [44]. The autocorrelation function in [41] is computed with an input signal in 6-bit precision; this does not reduce the number of operations, but faster and smaller multipliers or look-up tables can be used to speed up the computation and to reduce the chip size. Experiments have shown that the restoration results are not much influenced by this simplification.

For the calculation of $\hat{\mathbf{x}}$ in (2.13) it makes sense to analyze the matrix $\tilde{\mathbf{G}}$, defined in (3.8) in some detail. It can be seen that $\tilde{\mathbf{G}}$ is a symmetric matrix with constant values b_0 on its main diagonal. Furthermore, in Section 2.2 it was shown that $\tilde{\mathbf{G}}$ is positive definite.

The fact that $\tilde{\mathbf{G}}$ is positive definite allows one to use Cholesky decomposition [22,45] of $\tilde{\mathbf{G}}$ for solving $\hat{\mathbf{x}}$ from (2.13) in $O(m^3)$ operations. In the case of a burst of unknown samples $\tilde{\mathbf{G}}$ is Toeplitz, so that the system (2.13) can be solved more efficiently in $O(m^2)$ operations by the Levinson algorithm [46,22]. Even in the case of a more general pattern of unknown samples, $\tilde{\mathbf{G}}$ is related to a Toeplitz matrix, so that the system

(2.13) can be solved more efficiently than by Cholesky decomposition by using generalized Levinson algorithms [47]. However, this requires rather involved mathematics and does not lead to a less complicated hardware implementation since which generalized Levinson algorithm to use depends largely on the pattern of unknown samples. For these reasons only the solution of $\hat{x}$ from (2.13) by using Cholesky decomposition is considered.

In a Cholesky decomposition the matrix $\tilde{G}$ is decomposed into one of the following products:

$$\tilde{G} = LL^T, \tag{3.26}$$
$$\tilde{G} = \tilde{L}D\tilde{L}^T. \tag{3.27}$$

In (3.26), L is a lower triangular $m \times m$ matrix; in (3.27) $\tilde{L}$ is a lower triangular $m \times m$ matrix with constant values 1 on its main diagonal, and D is a diagonal $m \times m$ matrix with $d_{i,i} = l_{i,i}^2$, $i = 1, \ldots, m$. The systems

$$\tilde{G}\hat{x} = LL^T\hat{x} = -z$$

and

$$\tilde{G}\hat{x} = \tilde{L}D\tilde{L}^T\hat{x} = -z$$

are now solved by first solving y and $\tilde{y}$ from $Ly = -z$ and from $\tilde{L}\tilde{y} = -z$, respectively, and then by solving $\hat{x}$ from $L^T\hat{x} = y$ and $\tilde{L}^T\hat{x} = D^{-1}\tilde{y}$ respectively. Since the matrices L and $\tilde{L}$ are triangular, this can be done efficiently by back substitution.

Both forms of Cholesky decomposition take $O(m^3)$ operations. A drawback of the decomposition (3.26) is that it requires the computation of square roots. On the other hand, as is shown further on in equations (3.28)–(3.31), the elements of L in (3.26) satisfy bounds that are more convenient if an implementation in fixed point is envisaged.

For the elements of the matrices L and D, one can derive the results

$$1 \le l_{j,j} = \sqrt{d_{j,j}} \le \sqrt{b_0}, \ j = 1, \ldots, m, \tag{3.28}$$

and

$$\sum_{i=1}^{m} l_{i,j}^2 = b_0, \ j = 1, \ldots, m, \tag{3.29}$$

so that

$$|l_{i,j}| \le \sqrt{b_0 - 1}, \ i = 1, \ldots, j-1, \ j = 1, \ldots, m. \tag{3.30}$$

On substitution of $l_{i,j} = \tilde{l}_{i,j}\sqrt{d_{j,j}}$ into (3.30), and by using (3.28), one obtains

$$|\tilde{l}_{i,j}| \le \sqrt{b_0 - 1}, \ i = 1, \ldots, j-1, \ j = 1, \ldots, m. \tag{3.31}$$

The bounds (3.29), (3.30) and the right-hand bound of (3.28) can be derived by using results of [32, section 7] and by the fact that $\tilde{g}_{i,i} = b_0$, $i = 1, \ldots, m$. The left-hand bound is derived in Appendix F.

In a fixed-point implementation it is more convenient to solve the system

$$\tilde{G}'\hat{x} = -z',$$

where

$$\tilde{G}' = \frac{1}{b_0}\tilde{G}$$

and

$$z' = \frac{1}{b_0}z$$

because the absolute values of the elements of $\tilde{G}'$ are all bounded by 1. Then

$$\tilde{G}' = L'L'^T = \tilde{L}D'\tilde{L}^T,$$

where

$$L' = \frac{1}{\sqrt{b_0}}L, \, D' = \frac{1}{b_0}D.$$

On substituting this into (3.28), (3.29) and (3.30) one obtains

$$\frac{1}{\sqrt{b_0}} \leq l'_{j,j} = \sqrt{d'_{j,j}} \leq 1, \, j = 1, \ldots, m, \tag{3.32}$$

and

$$\sum_{i=1}^{m} l'^2_{i,j} = b_0, \, j = 1, \ldots, m, \tag{3.33}$$

so that

$$|l'_{i,j}| \leq 1, \, i = 1, \ldots, j - 1, \, j = 1, \ldots, m. \tag{3.34}$$

Now the $L'L'^T$ decomposition of $\tilde{G}'$ has the advantage over the $\tilde{L}D'\tilde{L}^T$ decomposition that the absolute values of the elements of L' are bounded by 1, and that the fixed-point operations can be performed without pre-scaling. The lower bound in (3.32) is important because the $l'_{j,j}$, $j = 1, \ldots, m$, are divisors in the process of back substitution and accuracy will be lost if they are too small. For digitized music, experiments have shown that b_0 has rather modest values, say $b_0 < 4$, so that the $l'_{j,j}$ do not become too small.

3.7 Autoregressive-moving average processes

The question may arise as to why in this book a sample restoration method is presented for autoregressive processes and not for the, more general, class of autoregressive-moving average processes. There are two important reasons that make the autoregressive-moving average model unattractive as a basis for a sample restoration method, both of them will be discussed in this section.

An autoregressive-moving average process

$$\underline{s}_k, \ k = -\infty, \ldots, +\infty,$$

is defined by the following expression:

$$\sum_{l=0}^{p} a_l \underline{s}_{k-l} = \sum_{l=0}^{q} c_l \underline{e}_{k-l}, \ k = -\infty, \ldots, +\infty. \tag{3.35}$$

In (3.35) q is the order of the moving average part of the process, the $c_0, \ldots, c_1$ are the *moving average coefficients*, and, as in (3.1), the $\underline{e}_k$, $k = -\infty, \ldots, +\infty$, is a zero-mean white noise process. An autoregressive-moving average process has a spectrum

$$S(\theta) = \frac{\left| \sum_{l=0}^{q} c_l \exp(-j\theta l) \right|^2}{\left| \sum_{l=0}^{p} a_l \exp(-j\theta l) \right|^2},$$

and it can be seen as the output of a filter with a transfer function

$$\frac{C(\exp(j\theta))}{A(\exp(j\theta))} = \frac{\sum_{l=0}^{q} c_l \exp(-j\theta l)}{\sum_{l=0}^{p} a_l \exp(-j\theta l)},$$

excited with zero-mean white noise with variance σ_e^2.

The first reason that the autoregressive-moving average model is unattractive is that estimation of the moving average coefficients is difficult. It is shown in Section 3.5 that estimating prediction coefficients of an autoregressive process is a quadratic problem with a unique solution. This is not the case with estimating moving average coefficients. Some iterative solutions exist [30] but convergence cannot be guaranteed. Furthermore, the computation of the moving average coefficients requires substantially more computations than the computation of the prediction coefficients of an autoregressive process.

The second reason that this model is unattractive is that the sequence

$$g_k = \int_{-\pi}^{\pi} \frac{\left| \sum_{l=0}^{p} a_l \exp(-j\theta l) \right|^2}{\left| \sum_{l=0}^{q} c_l \exp(-j\theta l) \right|^2} \exp(j\theta k) \, d\theta, \ k = -\infty, \ldots, +\infty,$$

that is required to build the system (2.13) is not of finite length. This means that either a truncated version of the sequence g_k must be used, introducing an error in the syndrome z in (2.13), or the matrix G must be computed from (2.15). The latter solution is unattractive from the point of view of computation.

A practical approach to the restoration of autoregressive-moving average processes is to assume that they can be modelled as an autoregressive process of a higher order.

3.8   Results

In this section the performance of the adaptive restoration method for autoregressive processes is considered for the following test signals:

1. **Artificially generated realizations of an autoregressive process of tenth order with a peaky spectrum.** Table 3.2 and Figure 3.4 show the prediction coefficients and the spectrum. Ten statistically independent sequences of 512 samples each were used. The excitation noise sequences were uncorrelated pseudo-random sequences with a Gaussian probability density function with zero mean and unit variance. The patterns of the unknown samples were bursts of lengths $m = 1, 4, 16, 50$.

2. **Artificially generated realizations of an autoregressive process of tenth order with a smooth spectrum.** Table 3.2 and Figure 3.4 show the prediction coefficients and the spectrum. The ten statistically independent sequences of 512 samples that were used were generated in the same manner as the test signals described under 1. The patterns of the unknown samples were bursts of lengths $m = 1, 4, 16, 50$.

3. **Multiple sinusoids.** A sequence of 512 samples, given by

$$s_n = 100 \sin(0.23\pi n + 0.3\pi) + 60 \sin(0.4\pi n + 0.3\pi) \qquad (3.36)$$

   was used. The patterns of the unknown samples were bursts of lengths $m = 4, 8, 14, 16$.

4. **Digital audio signals.** Bursts of 4, 6 or 16 unknown samples, occurring at a rate of 10 per second in a fragment of 36 seconds from a compact disc recording of Beethoven's Violin Concerto have been restored. The sample frequency of the signal was 44.1 kHz, so that a burst of 16 samples had a duration of 0.36 ms.

	Prediction coefficients	
	Peaky spectrum (1)	Smooth spectrum (2)
i	a_i	a_i
0	0.10000000000000D+01	0.10000000000000D+01
1	−0.84538931513857D+00	−0.71858091786779D+00
2	0.76158831819825D+00	0.55024755989824D+00
3	0.12977849626292D+00	0.79700219017466D−01
4	0.30387326386341D−02	0.15862374294460D−02
5	0.85694798229335D−01	0.38023237227971D−01
6	0.14042707569132D+00	0.52962003577618D−01
7	0.87115349165802D−01	0.27927184980177D−01
8	−0.42344306979864D−01	−0.11538422441359D−01
9	−0.23427477005693D+00	−0.54262006831786D−01
10	−0.30186789185306D−01	−0.59430061398161D−02

Table 3.2: Prediction coefficients of the autoregressive processes used as test signals 1 and 2 in this section.

5. **Digitized speech signals.** Bursts of 100 unknown samples, occurring at a rate of 10 per second in 10 English sentences of male and female speech have been restored. The sample frequency of the signal was 8 kHz, so that the bursts had a duration of 12.5 ms.

6. **Artificially generated realizations of an autoregressive process corrupted by white noise.** To the sequences described under 1 pseudo-random Gaussian white noise with zero mean was added. Signal-to-noise ratios of 40 and 20 dB were considered. The pattern of unknown samples was a burst of length 16.

7. **Sinusoids corrupted by white noise.** Pseudo-random Gaussian white noise was added to the sequences described under 3. Signal-to-noise ratios of 40 and 20 dB were considered. The patterns of the unknown samples were bursts of lengths $m = 4, 8, 14, 16$.

The test signals 1–7 were restored with the aid of the two following versions of the adaptive restoration method for autoregressive processes:

1. **The method using the autocovariance method to obtain the prediction coefficients.** This method is denoted by $C^{(i)}$, where i denotes the number of iterations.

2. **The method using the autocorrelation method to obtain the prediction coefficients.** This method is denoted by $R^{(i)}$, where i denotes the number of iterations.

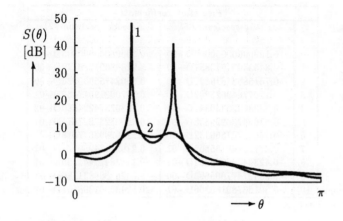

Figure 3.4: Peaky spectrum of test signal 1 and smooth spectrum of test signal 2.

The results are presented after one and after three iterations. For all test signals the performances of the adaptive restoration methods are judged by means of the relative quadratic restoration error $\hat{E}$:

$$\hat{E} = \frac{\frac{1}{m} \sum_{i=1}^{m} \left(\hat{s}_{t(i)} - s_{t(i)} \right)^2}{\frac{1}{N} \sum_{i=1}^{N} s_i^2}. \tag{3.37}$$

Since

$$\mathcal{E} \left\{ \sum_{i=1}^{m} \left(\hat{\underline{s}}_{t(i)} - \underline{s}_{t(i)} \right)^2 \right\} = \sigma_e^2 \, \text{trace} \left(\tilde{\mathbf{G}}^{-1} \right),$$

the relative quadratic restoration error $\hat{E}$ can be seen as an estimate for the relative restoration error variance per sample E defined in (3.11). The (averaged) value of $\hat{E}$ is presented for the test signals in Tables 3.3–3.11. Diagrams of some typical restoration results are presented in Figures 3.5–3.14, together with the original signals, in which the correct values of the unknown samples have been substituted. In the figures the original signal is marked by a (1), the restoration result by a (2), and the positions of the unknown samples indicated on the time axis. Besides the tables and the figures, the performances of the adaptive restoration method on both the music and the speech signals are also evaluated by listening tests.

For the test signals 1 and 2 the restoration results are compared with those obtained by using the true prediction coefficients. In the tables this method is denoted by F, where F stands for fixed coefficients. In the

m	p	N	F	$C^{(1)}$	$C^{(3)}$
1	10	512	0.79E$-$02	0.90E$-$02	0.72E$-$02
4	10	512	0.12E$-$01	0.11E$-$01	0.13E$-$01
16	10	512	0.26E$-$01	0.27E$-$01	0.26E$-$01
50	10	512	0.60E$-$01	0.61E$-$01	0.56E$-$01
m	p	N		$R^{(1)}$	$R^{(3)}$
1	10	512		0.90E$-$02	0.79E$-$02
4	10	512		0.13E$-$01	0.12E$-$01
16	10	512		0.28E$-$01	0.26E$-$01
50	10	512		0.62E$-$01	0.57E$-$01

Table 3.3: Average restoration errors with known coefficients and after 1 and 3 iterations for 10 realizations of an autoregressive process of order 10 with a peaky spectrum (test signal 1).

case of a single unknown sample, this non-adaptive restoration method amounts to the method presented in [16].

The tables and figures show restoration errors for various segment lengths and prediction orders. The true prediction orders of the artificially generated autoregressive processes and of the multiple sinusoids are known in advance: for the autoregressive processes it order is 10, for the multiple sinusoids it is twice the number of sinusoids in the signal, in this case $p = 4$. For these signals the true prediction order is used in most cases, though a higher prediction order is sometimes tried in order to achieve improvement in the restoration quality. For the music and the speech signal $p = \min(3m + 2, 50)$ is chosen. This rather arbitrary choice gives good results. The pattern of the unknown samples is always a burst. It usually happens that general patterns of unknown samples are restored with a smaller error than bursts with the same number of unknown samples.

The tables and figures give rise to the following remarks. From Tables 3.3 and 3.4 it is seen that the restoration errors for both adaptive methods do not differ significantly from the restoration errors for the restoration method that uses the true prediction coefficients. It seems that the estimation of the prediction coefficients from the incomplete data does not influence the quality of the restoration. It is also seen from Tables 3.3 and 3.4 that more than one iteration does not give a significant improvement. However, if the segment length N is smaller, more iterations do give an improvement, as can be seen from Tables 3.6 and 3.7. This phenomenon could also be observed in Example 4. Here results close to those of Tables 3.3 and 3.4 are obtained after three iterations. In general, restoration errors for autoregressive processes with a peaky spectrum are

m	p	N	F	$C^{(1)}$	$C^{(3)}$
1	10	512	0.31E+00	0.34E+00	0.32E+00
4	10	512	0.50E+00	0.62E+00	0.62E+00
16	10	512	0.99E+00	0.10E+01	0.10E+00
50	10	512	0.11E+01	0.11E+01	0.11E+01
m	p	N	F	$R^{(1)}$	$R^{(3)}$
1	10	512		0.34E+00	0.32E+00
4	10	512		0.62E+00	0.61E+00
16	10	512		0.10E+01	0.10E+01
50	10	512		0.11E+01	0.11E+01

Table 3.4: Average restoration errors with known coefficients and after 1 and 3 iterations for 10 realizations of an autoregressive process of order 10 with a smooth spectrum (test signal 2).

m	p	N	$C^{(1)}$	$C^{(3)}$	$R^{(1)}$	$R^{(3)}$
16	4	512	0.64E−02	0.28E−28	0.78E−01	0.16E−01
16	10	512	0.16E−06		0.30E−05	0.17E−06

Table 3.5: Average restoration errors after 1 and 3 iterations for a sum of 2 sinusoids (test signal 3). For $p = 10$, after 3 iterations the autocovariance matrix is almost singular.

m	p	N	$C^{(1)}$	$C^{(3)}$	$R^{(1)}$	$R^{(3)}$
16	10	64	0.42E−01	0.23E−01	0.63E−01	0.25E−01

Table 3.6: Average restoration errors after 1 and 3 iterations for 10 short realizations of an autoregressive process of order 10 with a peaky spectrum (test signal 1).

substantially smaller than for processes with a smooth spectrum.

For sinusoids the restoration error is theoretically zero. Indeed, Table 3.5 shows very small restoration errors for methods $C^{(1)}$ and $C^{(3)}$. The poorer results for $R^{(1)}$, $R^{(3)}$, $p = 4$ can be explained by the fact that the autocorrelation method uses a biased estimate for the autocorrelation function. This has less influence on the result if p is chosen to be greater than 4. If the autocovariance method is used to estimate the prediction coefficients, p must not be chosen much higher than the true order of prediction for after more than one iteration the autocovariance matrix will become almost singular and the prediction coefficients can no longer

m	p	N	$C^{(1)}$	$C^{(3)}$
16	4	64	0.52E+00	0.58E−20
m	p	N	$R^{(1)}$	$R^{(3)}$
16	10	64	0.56E−01	0.70E−03

Table 3.7: Average restoration errors after 1 and 3 iterations for a short sequence of a sum of 2 sinusoids (test signal 3).

m	p	N	$R^{(1)}$
4	14	128	0.98E−02
6	20	192	0.16E−01
16	50	512	0.19E−01

Table 3.8: Average restoration errors for a fragment of Beethoven's Violin Concerto (test signal 4).

m	p	N	$R^{(1)}$
100	50	512	0.56E+00

Table 3.9: Average restoration errors for 10 English sentences pronounced by a male and a female voice (test signal 5).

be calculated straightforwardly by solving the system (3.19). As can be seen from Tables 3.3, 3.4 and 3.11, for other signals than multiple sinusoids there are no significant differences between the restoration results obtained by using the autocovariance or the autocorrelation method.

For the music signal the relative quadratic restoration errors for the adaptive restoration methods are of the same orders of magnitude as those for the autoregressive processes with a peaky spectrum, which can be seen from Table 3.8. The high value for the relative quadratic restoration error for the speech signal in Table 3.9 can be explained as follows. In popular speech models [34], speech is assumed to consist of voiced parts - where the speech signal is highly periodic - and unvoiced parts - where the speech can be modelled as an autoregressive process of order approximately 12. In the voiced case, the speech spectrum contains many sharp, equidistant peaks, and the restoration results are similar to those obtained with autoregressive signals that have a peaky spectrum. In the unvoiced case, the speech spectrum is rather flat, and the restoration

SNR	m	p	N	F	$C^{(1)}$	$R^{(1)}$
40 dB	16	10	512	0.26E−01	0.27E−01	0.28E−01
40 dB	16	20	512		0.28E−01	0.30E−01
20 dB	16	10	512	0.42E−01	0.42E−01	0.43E−01
20 dB	16	20	512		0.45E−01	0.46E−01

Table 3.10: Average restoration errors for various signal-to-noise ratios and orders of prediction with known coefficients after 1 iteration for 10 realizations of an autoregressive process of order 10 with a peaky spectrum (test signal 6).

results are similar to those obtained with autoregressive signals with a smooth spectrum. As can be seen from Table 3.4 and Figure 3.7, these results are rather poor, especially if the bursts are large. The relative quadratic restoration error in Table 3.9 is averaged over 20 sentences and the high value is caused by the presence of unvoiced and silent fragments. However, the poor restoration results for unvoiced and silent fragments do not cause any audible disturbance in the restored speech. Figure 3.14 shows a typical restoration result for voiced speech.

Listening tests have revealed that the restoration errors in the test signals 4, as well as in many other signals, are practically inaudible. After increasing the burst length from 16 to 50 the restoration results are still quite good for most music signals, although some restoration errors become audible. For the speech signals, bursts can be restored up to 100 unknown samples without audible errors. In the case of speech signals it may seem curious that the method still works for bursts of these lengths (which represent time intervals of durations up to 12.5 ms), since the length N of the segment used to estimate the prediction coefficients represents a time interval of more than 60 ms which is generally too long for a speech signal to be assumed stationary. However, some speech sounds, for instance vowels, can be assumed stationary for several hundreds of milliseconds, and for these the method performs well. Other speech sounds, especially the plosive sounds, /b/, /d/, /g/, /p/, /t/ and /k/, can only be assumed stationary for a few milliseconds and cannot be restored correctly. Still, the errors made here do not seem to reduce the subjective restoration quality, possibly because of masking effects.

With a decreasing signal-to-noise ratio the restoration results of the adaptive restoration methods deteriorate slightly. However, they still do not differ significantly from the results that could be obtained if the true prediction coefficients were used. This can be seen from Tables 3.10 and 3.11.

SNR	m	p	N	$C^{(1)}$	$C^{(3)}$	$R^{(1)}$	$R^{(3)}$
40 dB	16	4	512	0.84E−02	0.40E−03	0.87E−01	0.21E−01
40 dB	16	10	512	0.12E−03	0.11E−03	0.12E−03	0.12E−03
20 dB	16	4	512	0.38E+00	0.33E+00	0.44E+00	0.40E+00
20 dB	16	10	512	0.10E−01	0.11E−01	0.10E−01	0.10E−01

Table 3.11: Average restoration errors for various signal-to-noise ratios and orders of prediction for a sum of sinusoids (test signal 7).

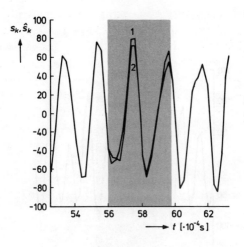

Figure 3.5: Original (1) and restored (2) autoregressive process with a peaky spectrum, $m = 16$, $p = 10$, $N = 512$, autocovariance method, 1 iteration, $\hat{E} = 0.23\text{E}{-}01$.

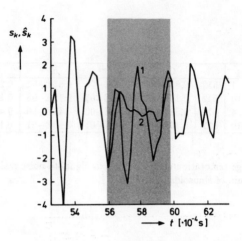

Figure 3.6: Original (1) and restored (2) autoregressive process with a smooth spectrum, $m = 16$, $p = 10$, $N = 512$, autocovariance method, 1 iteration, $\hat{E} = 0.10\text{E}+01$.

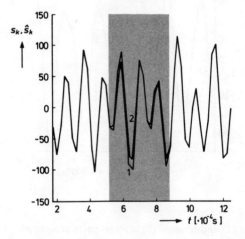

Figure 3.7: Original (1) and restored (2) autoregressive process with a peaky spectrum, $m = 16$, $p = 10$, $N = 64$, autocovariance method, 1 iteration, $\hat{E} = 0.38\text{E}-01$.

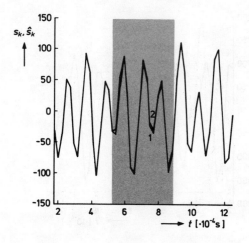

Figure 3.8: Original (1) and restored (2) autoregressive process with a peaky spectrum, $m = 16$, $p = 10$, $N = 64$, autocovariance method, 3 iterations, $\hat{E} = 0.12\mathrm{E}-01$.

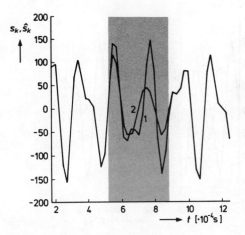

Figure 3.9: Original (1) and restored (2) test signal 3, $m = 16$, $p = 10$, $N = 64$, autocovariance method, 1 iteration, $\hat{E} = 0.52\mathrm{E}+00$.

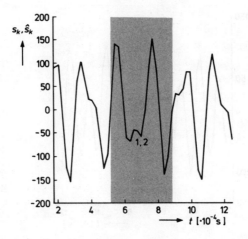

Figure 3.10: Original (1) and restored (2) test signal 3, $m = 16$, $p = 10$, $N = 64$, autocovariance method, 3 iterations, $\hat{E}$ =0.58E−02.

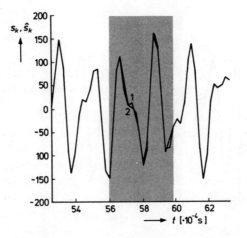

Figure 3.11: Original (1) and restored (2) signal 7, signal-to-noise ratio (SNR) =20 dB, $m = 16$, $p = 10$, $N = 512$, autocovariance method, 1 iteration, $\hat{E}$ =0.10E−01.

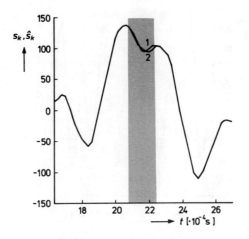

Figure 3.12: Original (1) and restored (2) music signal, $m = 6$, $p = 20$, $N = 192$, autocorrelation method, 1 iteration, $\hat{E} = 0.59\mathrm{E}{-}02$.

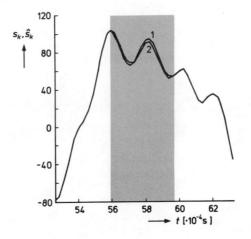

Figure 3.13: Original (1) and restored (2) music signal, $m = 16$, $p = 50$, $N = 512$, autocorrelation method, 1 iteration, $\hat{E} = 0.22\mathrm{E}{-}01$.

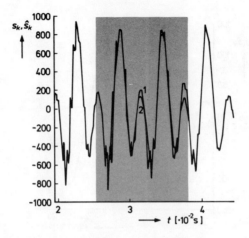

Figure 3.14: Original (1) and restored (2) speech signal, $m = 100$, $p = 50$, $N = 512$, autocorrelation method, 1 iteration, $\hat{E} = 0.70\mathrm{E}{-}01$.

Chapter 4

Speech signals

4.1 Introduction

In this chapter an adaptive restoration method for unknown samples in a digitized speech signal is derived.[1] The problem of unknown samples in speech signals occurs in car telephony systems, known as mobile automatic telephony systems, where dips may occur in the received mobile automatic telephony signal. The reason is that the car drives through a pattern of standing waves and when passing through a node no signal is received for a short period of time. The length of this period depends on the speed of the car. During these dips the speech is heavily distorted. If the speech signal is sampled, bursts of samples are erroneous. Their positions can be derived from the received carrier signal, which also contains a dip. The erroneous samples are considered to be unknown and have to be restored. The burst length t_{error} and the length of the error-free intervals $t_{correct}$ depend on the vehicle speed v. Figure 4.1[2] shows this dependence for a speech signal that has been frequency modulated on a carrier frequency of 300 MHz. The sampling frequency of digitized speech is 8 kHz. From Figure 4.1 it follows that the burst length m varies between about 8 at a speed of 100 km/h and about 100 at a speed of 3 km/h.

From the results presented in Section 3.8 it is clear that the restoration method for autoregressive processes presented in Chapter 3 is very well capable of restoring large patterns of unknown samples in speech signals. A problem is, however, that for these large patterns the number of operations that is required to solve the system (2.13) becomes very

---

[1]This chapter has been published previously in [10].

[2]Figure 4.1 is derived from an original produced by Dr L. Kittel, who was at that time with Philips Kommunikations Industrie, Nürnberg.

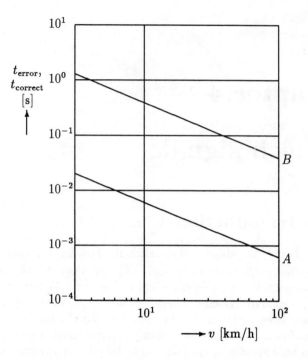

Figure 4.1: Vehicle speed versus burst length (A) and vehicle speed versus length of error-free interval (B).

high. Assuming that the error pattern is a burst, the number of operations required is $O(m^2)$. This is high considering that in reality m can be as high as 100, and present-day hardware is not capable of solving the system in real time at the required rate. For this reason another algorithm has been developed for sample restoration in speech signals. This algorithm tries to exploit the specific properties of speech. Its main assumption is that speech is often *quasi-periodic*. As is the case with the restoration method for autoregressive processes, a function is minimized with respect to the unknown samples, but in this case the method is not iterative. The unknown samples are solved from a linear system of equations in $O(m)$ operations. The results obtained with this method are of the same quality as those obtained with the restoration method for autoregressive processes, but the required hardware is substantially less complex.

This chapter is organized as follows. In Section 4.2 the properties of the speech signal that are of importance for the restoration problem are discussed. The algorithm is derived in Section 4.3 and implementational aspects are discussed in Section 4.4. Restoration results are shown in Section 4.5

4.2   The speech signal

Speech is often modelled as the output of a time-varying all-pole filter. This filter is excited either by white noise, in which case the speech is called *unvoiced*, or by an approximately periodic sequence of pulses, in which case the speech is called *voiced* [34,9]. The 'period' of this sequence is called the pitch period, T_p. Its duration is between 2 and 20 ms, which, at a sampling frequency of 8 kHz, corresponds to a number of samples q that is between 16 and 160. The order p of the all-pole filter is usually chosen somewhere between 10 and 16. Unvoiced speech, according to this model, is an autoregressive process. Figure 4.2 gives a diagram of the speech production system according to this model. Figure 3.1 shows an example of an all-pole filter of order $p = 3$. Figures 4.6 and 4.8 of Section 4.5 show examples of voiced speech, Figure 4.10 shows an example of unvoiced speech.

The use of this model may be justified as follows. The all-pole filter models the transfer function of the vocal tract, consisting of the cavities through which the sound is passed [9]. If during speech production the vocal chords are brought into resonance, the pressure at the beginning of the vocal tract varies pulse-wise and the speech is voiced. This is the case for all the vowels and some of the consonants, for example /b/, /m/, /n/, /z/. If the air passes the vocal chords without causing them to resonate, the air flow at the beginning of the vocal tract is turbulent; one could say that the pressure variations resemble white noise, and the speech is unvoiced. This is the case for example with the consonants /s/ and /f/. The plosive sounds, such as /t/ and /p/, are not incorporated in this model. The model is often used in speech coding [34].

4.3   The restoration method

The speech model of Section 4.2 is still too complicated to be used as a basis for a sample restoration algorithm that can be implemented in real-time operating hardware. According to this model two restoration methods are required: one for unvoiced speech (this can be the sample restoration method for autoregressive processes) and one for voiced speech. In addition, before every restoration a voiced/unvoiced decision

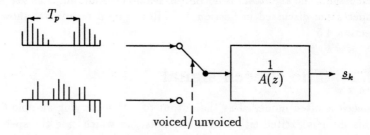

Figure 4.2: Model of the speech production system.

has to be taken. In general, this decision is not reliable. Experiments have shown that the penalty for restoring voiced speech as if it were unvoiced is high, whereas the penalty for restoring unvoiced speech as if it were voiced is minimal. Therefore, the first simplification is that the signal is always assumed to be voiced. The choice of a sample restoration method comes down to the choice of a matrix $\mathbf{G}$, as defined in Section 2.2. In this case $\mathbf{G}$ depends on the coefficients a_i, $i = 0, \ldots, p$, of the all-pole filter and the autocorrelation function of the sequence of almost-periodic pulses that excite the all-pole filter. A $\mathbf{G}$ computed in this way would be similar to the matrix $\mathbf{G}$ for autoregressive processes, and therefore would not lead to a less complex implementation. For that reason, a second simplification is introduced by ignoring the transfer function $1/A(z)$ and assuming that the speech signal is *quasi-periodic*. In this case this means that the sample $\underline{s}_k$ resembles closely the sample $\underline{s}_{k-q}$, where q is the pitch period, expressed in sample distances. This can be expressed by

$$\underline{s}_k - c\underline{s}_{k-q} = \underline{e}_k, \quad k = q + 1, \ldots, N. \tag{4.1}$$

Here c is called the *pitch coefficient*, for which it is assumed that

$$0 \le c < 1, \tag{4.2}$$

and the $\underline{e}_k$ are zero-mean white noise samples. In (4.1) it is assumed that the pitch period does not change over the available segment of data s_k, $k = 1, \ldots, N$. The third simplification is that only bursts of unknown samples are considered. For the application in mobile automatic telephony this is not really a restriction. It is assumed that for every value of q

$$q + 1 \le t(1), \ t(m) = t(1) + m - 1 \le N - q. \tag{4.3}$$

Since the pitch period q varies between $q_{min} = 16$ and $q_{max} = 160$, this implies that

$$q_{max} + 1 \leq t(1), \ t(m) = t(1) + m - 1 \leq N - q_{max}. \qquad (4.4)$$

This means that there is a guard space of at least $q_{max} = 160$ samples on either side of the burst.

The stochastic signal $\underline{s}_k$ in (4.1) has a spectrum

$$S(\theta) = \frac{\sigma_e^2}{-c\exp(j\theta q) + (1 + c^2) - c\exp(-j\theta q)}, \qquad (4.5)$$

so that if q and c are known the sequence g_k (2.21) can be defined by

$$g_k = \left\{ \begin{array}{ll} 1, & k = 0, \\ \alpha = -\frac{c}{1+c^2}, & |k| = q, \\ 0, & \text{otherwise.} \end{array} \right. \qquad (4.6)$$

In this manner the vector $\hat{\mathbf{x}}$ of unknown samples can be solved from a system (2.13)

$$\tilde{\mathbf{G}}\hat{\mathbf{x}} = -\mathbf{z},$$

with the $m \times m$ matrix $\tilde{\mathbf{G}}$ defined by

$$\tilde{g}_{i,j} = \left\{ \begin{array}{ll} 1, & i = j, \\ \alpha, & |i - j| = q, \\ 0, & \text{otherwise,} \end{array} \right. \qquad (4.7)$$

and the syndrome $\mathbf{z}$ defined by

$$z_i = \alpha(v_{t(i)-q} + v_{t(i)+q}), \ i = 1, \ldots, m. \qquad (4.8)$$

From the definition of α in (4.6) it follows that $-\frac{1}{2} < \alpha \leq 0$. Note that $\tilde{\mathbf{G}}$ is the identity matrix if $m \leq q$. In that case each estimate for an unknown sample is found as a weighted sum of two known samples.

The same result can be obtained by minimizing as a function of $\hat{\mathbf{x}}$ the function

$$Q_q(c, \mathbf{x}) = \sum_{k=q+1}^{N} \left(s_k - c s_{k-q} \right)^2. \qquad (4.9)$$

This function resembles the function $Q(\mathbf{a}, \mathbf{x})$ of (3.13). The pitch coefficient c is in fact a prediction coefficient, and just as $Q(\mathbf{a}, \mathbf{x})$ is quadratic in $\mathbf{a}$ for known $\mathbf{x}$ and quadratic in $\mathbf{x}$ for known $\mathbf{a}$, $Q_q(c, \mathbf{x})$ is quadratic in c for known $\mathbf{x}$ and quadratic in $\mathbf{x}$ for known c.

For the estimation of the pitch period q and the pitch coefficient c, many methods have been proposed in the literature, some of them in

[34]. A popular one is to minimize $Q_q(c, \mathbf{x})$ for known $\mathbf{x}$ as a function of q and c. This implies that for every q in the range $q = 16, \ldots, 160$, $Q_q(c, \mathbf{x})$ is minimized as a function of c. It can be shown that this is equivalent to finding the q for which the sequence

$$\rho_s(q) = \frac{\sum_{k=q_{max}+1}^{N} s_k s_{k-q}}{\sum_{k=q_{max}+1}^{N} s_k^2}, \quad q = 16, \ldots, 160 \qquad (4.10)$$

is maximal. This sequence is often called the *normalized correlation function*. The pitch coefficient is then found as

$$\hat{c} = \rho_s(\hat{q}). \qquad (4.11)$$

Since the sequence s_k, $k = 1, \ldots, N$, contains unknown samples, q and c cannot be estimated in this way. A solution might be to minimize $Q_q(c, \mathbf{x})$ iteratively as a function of q, c and $\mathbf{x}$, in much the same way as in Chapter 3, but this would complicate the method too much. Experiments have shown two important phenomena which allow further simplifications. The first is that the position $\hat{q}$ of the maximum $\rho_s(\hat{q})$ in the sequence $\rho_s(q)$, $q = 16, \ldots, 160$ does not change significantly if the sequence contains unknown samples that are set equal to zero. This implies that q can be estimated in the presence of unknown samples. The second phenomenon is that the restoration result is not sensitive to changes in c as long as c is close to 1. Therefore, c does not have to be estimated, but can be fixed in advance at a fairly high value, say $c = 0.9$. A value of c very close to 1 sometimes leads to audible clicks. These occur if q has been estimated incorrectly, which does occasionally happen. A fixed value for c leads to a simpler pitch estimation, as is shown in the next section, and also to a simple method of solving the system (2.13), as is shown in Section 4.4.

Estimation of q is in fact the estimation of the periodicity present in the signal. This is not affected by some non-linear instantaneous transformations on the signal. For instance, instead of the signal s_k a clipped version of the signal, defined by

$$\bar{s}_k = \begin{cases} +1, & s_k > t_s, \\ 0, & -t_s \le s_k \le t_s, \\ -1, & s_k < -t_s, \end{cases} \qquad (4.12)$$

can be used. Here t_s is a positive threshold which may be taken to be dependent on the samples s_k, $k = 1, \ldots, N$, for instance $t_s = 0.5|s_k|_{max}$. The computation of the sequence $\rho_{\bar{s}}(q)$, $q = 16, \ldots, 160$, requires simpler hardware than the computation of the sequence $\rho_s(q)$, $q = 16, \ldots, 160$,

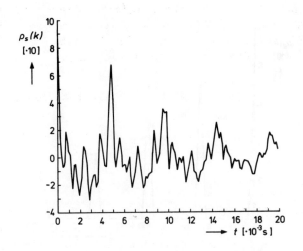

Figure 4.3: Normalized correlation function obtained with the original data sequence.

because no multipliers are needed. A disadvantage is that the $\rho_{\bar{s}}(\hat{q})$ obtained in this way is no longer a good estimate for c, but this is no problem here since c can be fixed in advance. With a well-chosen t_s, for instance $t_s = 0.5|s_k|_{\max}$ works well, the sequence $\rho_{\bar{s}}(k)$, $k = 16, \ldots, 160$, has sharper peaks than (4.10) and therefore sometimes gives even better estimates for q. This is illustrated in Figures 4.3 and 4.4, where the sequences $\rho_s(k)$ and $\rho_{\bar{s}}(k)$ are given for the data sequence of Figure 4.6 of Section 4.5.

4.4 Efficient implementation

To find the estimates for the unknown samples the pitch period q has to be estimated and the system (2.13) has to be built and solved. In Section 4.3 it is shown that by using clipped data the pitch estimation can be made efficient for computation. In this section the building of the system (2.13) and the method of solving it are considered.

Since the pitch coefficient c has been taken as fixed, α is also known in advance. The computation of the right-hand side of (2.13) by using (4.8) is already simple and need not be further simplified. The system in this case is solved by a well-known technique called LU decomposition [22], in which $\tilde{G}$ is written as a product $\mathbf{LU}$, where $\mathbf{L}$ is a lower triangular matrix containing 1s on its main diagonal, and $\mathbf{U}$ is an upper triangular

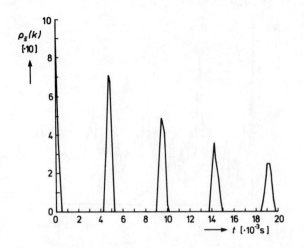

Figure 4.4: Normalized correlation function obtained with the clipped data sequence.

matrix. It is shown that, because α is known in advance, the matrix $\tilde{\mathbf{G}}$ does not have to be computed explicitly and that the unknown samples can be solved in $O(m)$ operations. The system

$$\tilde{\mathbf{G}}\hat{\mathbf{x}} = \mathbf{LU}\hat{\mathbf{x}} = -\mathbf{z}$$

is solved by first solving $\mathbf{y}$, using back substitution, from

$$\mathbf{Ly} = -\mathbf{z}$$

and then by solving $\hat{\mathbf{x}}$, again by using back substitution, from

$$\mathbf{U}\hat{\mathbf{x}} = \mathbf{y}.$$

For the matrices $\mathbf{L}$ and $\mathbf{U}$ the recurrence relations

$$u_{i,j} = \begin{cases} 1, & j = i = 1, \ldots, q, \\ 1 - \dfrac{\alpha^2}{u_{i-q,i-q}}, & j = i = q+1, \ldots, m, \\ \alpha, & j = i+q,\ i = 1, \ldots, m-q, \\ 0, & \text{otherwise} \end{cases} \tag{4.13}$$

and

$$l_{i,j} = \begin{cases} 1, & j = i = 1, \ldots, m, \\ \dfrac{\alpha}{u_{i-q,i-q}}, & j = i-q,\ i = q+1, \ldots, m, \\ 0, & \text{otherwise} \end{cases} \tag{4.14}$$

can be derived. The matrices $\mathbf{L}$ and $\mathbf{U}$ are sparse, as is $\tilde{\mathbf{G}}$.

Example 5 *For the case of $m = 7$ and $q = 3$, the matrices $\tilde{G}$, L and U are given by*

$$\tilde{G} = \begin{bmatrix} 1 & 0 & 0 & \alpha & 0 & 0 & 0 \\ 0 & 1 & 0 & 0 & \alpha & 0 & 0 \\ 0 & 0 & 1 & 0 & 0 & \alpha & 0 \\ \alpha & 0 & 0 & 1 & 0 & 0 & \alpha \\ 0 & \alpha & 0 & 0 & 1 & 0 & 0 \\ 0 & 0 & \alpha & 0 & 0 & 1 & 0 \\ 0 & 0 & 0 & \alpha & 0 & 0 & 1 \end{bmatrix},$$

$$L = \begin{bmatrix} 1 & 0 & 0 & 0 & 0 & 0 & 0 \\ 0 & 1 & 0 & 0 & 0 & 0 & 0 \\ 0 & 0 & 1 & 0 & 0 & 0 & 0 \\ \alpha & 0 & 0 & 1 & 0 & 0 & 0 \\ 0 & \alpha & 0 & 0 & 1 & 0 & 0 \\ 0 & 0 & \alpha & 0 & 0 & 1 & 0 \\ 0 & 0 & 0 & \frac{\alpha}{1-\alpha^2} & 0 & 0 & 1 \end{bmatrix},$$

$$U = \begin{bmatrix} 1 & 0 & 0 & \alpha & 0 & 0 & 0 \\ 0 & 1 & 0 & 0 & \alpha & 0 & 0 \\ 0 & 0 & 1 & 0 & 0 & \alpha & 0 \\ 0 & 0 & 0 & 1-\alpha^2 & 0 & 0 & \alpha \\ 0 & 0 & 0 & 0 & 1-\alpha^2 & 0 & 0 \\ 0 & 0 & 0 & 0 & 0 & 1-\alpha^2 & 0 \\ 0 & 0 & 0 & 0 & 0 & 0 & \frac{1-2\alpha^2}{1-\alpha^2} \end{bmatrix}.$$

The LU decomposition of $\tilde{G}$ is related to the decomposition $\tilde{L}D'\tilde{L}^T$ of Section 3.6. In fact, one has

$$L = \tilde{L}$$

and

$$U = D'\tilde{L}^T.$$

From these expressions, and by using (3.32), (3.34) and the fact that $0 \le c < 1$, it follows that

$$u_{i,i} > \frac{1}{2} \qquad (4.15)$$

and that

$$|l_{i,i-q}| \le 1. \qquad (4.16)$$

The $u_{i,i}$ are divisors, and from (4.15) it follows that they do not become too small.

The fact that both L and U are sparse simplifies the back substitution. The estimates for the unknown samples are calculated by evaluating the expressions

$$y_i = \begin{cases} z_i, & i = 1, \dots, q, \\ z_i - l_{i,i-q}y_{i-q}, & i = q + 1, \dots, m, \end{cases}$$

and

$$\hat{x}_{m+1-i} = \begin{cases} \dfrac{1}{u_{m+1-i,m+1-i}}y_{m+1-i}, & i = 1, \dots, q, \\ \dfrac{1}{u_{m+1-i,m+1-i}}(y_{m+1-i} - \alpha\hat{x}_{m+1+q-i}), & i = q + 1, \dots, m. \end{cases}$$

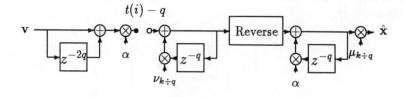

Figure 4.5: Sample restoration filter.

To compute $\mathbf{y}$ and $\hat{\mathbf{x}}$ it is convenient to store the sequences

$$\mu_k = \begin{cases} 1, & k = 0, \\ \frac{1}{1-\alpha^2\mu_{k-1}}, & k = 1, \ldots, m_{\max} \div q_{\min}, \end{cases}$$

and

$$\nu_k = \begin{cases} 0, & k = 0, \\ \alpha\mu_{k-1}, & k = 1, \ldots, m_{\max} \div q_{\min}. \end{cases}$$

Here $m_{\max}$ denotes the maximum allowed burst length and the symbol '$\div$' denotes integer division. The vectors $\mathbf{y}$ and $\hat{\mathbf{x}}$ are now computed by

$$y_i = \begin{cases} z_i, & i = 1, \ldots, q, \\ z_i - \nu_{(i-1)\div q} y_{i-q}, & i = q+1, \ldots, m, \end{cases}$$

and by

$$\hat{x}_{m+1-i} = \begin{cases} \mu_{(m-i)\div q} y_{m+1-i}, & i = 1, \ldots, q, \\ \mu_{(m-i)\div q}(y_{m+1-i} - \alpha\hat{x}_{m+1+q-i}), & i = q+1, \ldots, m. \end{cases}$$

The number of operations required to compute $\hat{\mathbf{x}}$ is $O(m)$. This process of back substitution can be visualized as follows. The syndrome $\mathbf{z}$ is taken as the input sequence for a recursive digital filter in zero state with one time-varying coefficient $\nu_{k\div q}$. The first m output samples of this filter are reversed in time and used as the input of another recursive filter in zero state with a time-varying gain factor $\mu_{k\div q}$. The first m output samples of this filter are the estimates for the unknown samples. The filter structure is shown in Figure 4.5. The input is the sequence $\mathbf{v}$, the first part computes the syndrome. The switch only closes at instants $t(1) - q, \ldots, t(m) - q$.

4.5   Results

From listening tests it can be concluded that in most cases the restoration method is capable of restoring bursts of up to 100 unknown samples in male and female speech. Compared with the results obtained on the same material with the method for autoregressive processes of Chapter 3 hardly any degradation can be heard. Inspection of the restored signals shows that in the rare cases where audible errors do occur, these are caused by an incorrect pitch estimate. Some typical restoration results are given in Figures 4.6–4.11. Figures 4.6, 4.8 and 4.10 show segments of $N = 512$ samples of the original signals, without errors. The first two are examples of voiced speech, the third is unvoiced. In these signals, bursts of $m = 100$ samples, located at the centre, were distorted and restored again. The pitch coefficient was fixed at $c = 0.9$, the threshold for clipping t_s was set to $t_s = 0.5|s_k|_{\max}$. The restored signals are shown in Figures 4.7, 4.9 and 4.11, respectively. The positions of the restored samples are marked on the time axis. It can be observed from Figure 4.11 that a 'noisy periodic' restoration was found, with a period of approximately 2.8 ms. This corresponds to the estimated pitch period of 2.75 ms. This effect occurs because the signal is assumed to be periodic, and therefore a periodic restoration is obtained. This effect was not audible in listening tests.

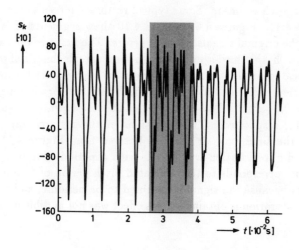

Figure 4.6: Original voiced speech segment.

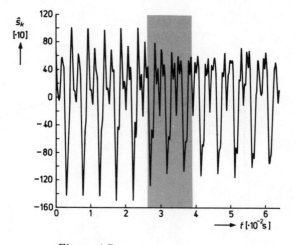

Figure 4.7: Restored voiced speech segment.

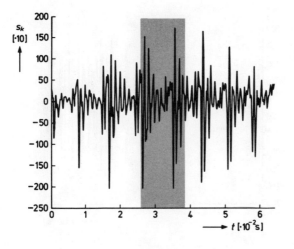

Figure 4.8: Original voiced speech segment.

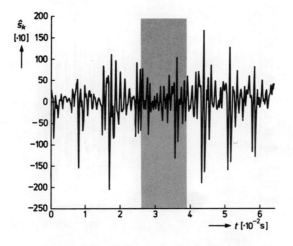

Figure 4.9: Restored voiced speech segment.

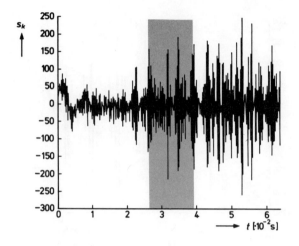

Figure 4.10: Original unvoiced speech segment.

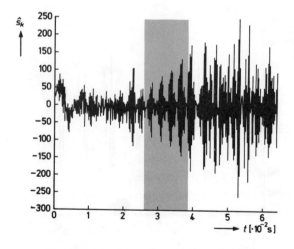

Figure 4.11: Restored unvoiced speech segment.

Chapter 5

Band-limited signals

5.1 Introduction

This chapter describes and analyzes a sample restoration method for band-limited signals that has been treated previously in [5,6]. The reason for its inclusion in this book is that it can be seen as a special case of the general sample restoration method of Chapter 2. The method is derived in Section 5.2, and an analysis of its numerical robustness is given in Section 5.3. It so happens that even though this method is theoretically capable of error-free sample restoration, it is so sensitive to numerical (rounding) errors and to deviations from the original assumption of band-limitedness that it is of no practical value if the product of bandwidth and the number of unknown samples is much greater than 1. This is demonstrated in Section 5.4 by some examples.

5.2 The restoration method

A signal is called band-limited if the signal spectrum $S(\theta)$ is identical with 0 on one or more finite subintervals of $[0, \pi]$. It is shown in Appendix A that in that case the infinite autocorrelation matrix is singular and that for N large enough the $N \times N$ autocorrelation matrix can be considered to be singular. First assume that an infinite segment of samples s_k, $k = -\infty, \ldots, +\infty$, is available and that the samples at positions $t(i)$, $i = 1, \ldots, m$ are unknown. A sequence g_k, $k = -\infty, \ldots, \infty$, with a Fourier transform $G(\exp(j\theta))$ has to be found such that (2.23)

$$\frac{1}{2\pi} \int_{-\pi}^{\pi} G(\exp(j\theta)) S(\theta) d\theta = 0.$$

71

By applying (2.24) and (2.25) the system (2.13)

$$\tilde{G}\hat{x} = -z$$

can be built and solved for the unknown samples. The $G(\exp(j\theta))$ chosen in [14,6,5] is such that it is 0 on every subinterval of $[0, \pi]$ where $S(\theta) > 0$, and it is identical with 1 on at least one subinterval of $[0, \pi]$, so that $G(\exp(j\theta))$ is the frequency transfer function of an ideal band-pass filter. In practice g_k, $k = -\infty, \ldots, +\infty$, is approximated by a finite sequence. Several methods exist to achieve this [48,49]. A simple method is to truncate the infinite sequence s_k and multiply it by a window function; it is also possible to find a finite sequence g_k by using more advanced design methods for selective filters. The restoration results do not depend much on the approximation method chosen to obtain a finite sequence g_k, $k = -p, \ldots, p$, and to simplify the analysis here g_k, $k = -p, \ldots, p$, is a truncated version of an ideal band-pass filter. Furthermore, it assumed that the signal is low-pass, as is the case in [14,5,6]. Similar derivations can be made for high-pass and band-pass signals. The coefficients g_k then follow from

$$g_k = \delta_k - \frac{\sin(\alpha k\pi)}{k\pi}, \quad k = -p, \ldots, p. \qquad (5.1)$$

Here α, $0 < \alpha < 1$, is the signal bandwidth, relative to half the sampling frequency. In [6] the influence of windowing the g_k obtained in this way is analyzed.

The $\tilde{G}$ derived from the g_k of (5.1) is positive definite, because

$$G(\exp(j\theta)) \geq 0.$$

This is explained in Appendix A. Its eigenvalues are in the open interval $(0, 1)$, as follows easily from the fact that both $\tilde{G}$ and $I - \tilde{G}$ are positive definite. As is shown in Chapter 2, sample restoration methods for signals with a singular autocorrelation matrix give theoretically a perfect restoration. It is shown, however, in Section 5.4 that the restoration method for band-limited signals of this chapter is not at all a good restoration method from a robustness point of view. The analysis of Section 5.3 shows that this method is very sensitive to out-of-band components, such as white noise, in the signal and that it is also sensitive to inaccuracies in the syndrome z.

5.3   Numerical robustness

The analysis of this section is made for a burst of unknown samples, which is the most critical case. Better results are obtained for more scattered patterns of unknown samples.

The numerical errors in the system (2.13) occur in $\mathbf{z}$, not in $\tilde{\mathbf{G}}$, which is known in advance, and can be stored in the required precision. First assume that, due to rounding errors, instead of $\mathbf{z}$ a vector $\mathbf{z}'$ has been obtained. By using the results of [22, section 2.5], it can be shown that

$$\frac{\| \hat{\mathbf{x}} - \mathbf{x} \|}{\| \mathbf{x} \|} \leq \frac{\lambda_{\max}}{\lambda_{\min}} \frac{\| \mathbf{z}' - \mathbf{z} \|}{\| \mathbf{z} \|} \cong \frac{1}{\lambda_{\min}} \frac{\| \mathbf{z}' - \mathbf{z} \|}{\| \mathbf{z} \|}. \tag{5.2}$$

Here, $\lambda_{\max}$ and $\lambda_{\min}$ are the maximum and minimum eigenvalues of $\tilde{\mathbf{G}}$, the latter approximation is a consequence of the fact that $\lambda_{\max} \cong 1$. In [50,6] it is derived that

$$\lambda_{\min} \cong \text{constant } \exp(-\alpha m \pi).$$

This shows that when the product αm increases, the influence of rounding errors in $\mathbf{z}$ increases exponentially.

A more detailed statistical analysis can be made if it is assumed that the segment s_k, $k = 1, \ldots, N$, is corrupted by white noise. It is then possible to analyze the influence of the out-of-band noise. The in-band noise is not taken into account because it cannot be distinguished from the signal. Assume that the signal is corrupted by white noise samples

$$\underline{n}_k = \underline{n}_{1,k} + \underline{n}_{2,k},$$

where $\underline{n}_{1,k}$ is the in-band component of the noise and $\underline{n}_{2,k}$ is the out-of-band component. If the variance of the white noise is σ_n^2, then the out-of-band component of the noise has variance $(1 - \alpha)\sigma_n^2$. Let

$$\underline{\mathbf{n}}_2 = [\underline{n}_{2,t(1)}, \ldots, \underline{n}_{2,t(m)}]^T.$$

It can be shown that

$$\underline{\mathbf{e}} = \hat{\underline{\mathbf{x}}} - \underline{\mathbf{x}} = -\tilde{\mathbf{G}}^{-1}(\mathbf{I} - \tilde{\mathbf{G}})\underline{\mathbf{n}}_2.$$

The error covariance matrix is given by

$$
\begin{aligned}
\mathbf{E} &= \mathcal{E}\{\underline{\mathbf{e}}\,\underline{\mathbf{e}}^T\} \\
&= (\tilde{\mathbf{G}}^{-1} - \mathbf{I})\mathcal{E}\{\underline{\mathbf{n}}_2\underline{\mathbf{n}}_2^T\}(\tilde{\mathbf{G}}^{-1} - \mathbf{I})^T \\
&= (1 - \alpha)\sigma_n^2\,(\mathbf{I} - \tilde{\mathbf{G}})\tilde{\mathbf{G}}^{-1}(\mathbf{I} - \tilde{\mathbf{G}})^T
\end{aligned}
$$

$$= (1 - \alpha)\sigma_n^2\,\mathbf{U} \begin{bmatrix} \frac{(1-\lambda_1)^2}{\lambda_1} & & \mathbf{0} \\ & \ddots & \\ \mathbf{0} & & \frac{(1-\lambda_m)^2}{\lambda_m} \end{bmatrix} \mathbf{U}^T. \tag{5.3}$$

Here

$$\mathbf{U} = [\mathbf{u}_1, \ldots, \mathbf{u}_m]^T$$

is an $m \times m$ matrix of which the columns are the eigenvectors $\mathbf{u}_i$ of the matrix $\tilde{\mathbf{G}}$ associated with the eigenvalues λ_i. It is of interest to analyze the spectrum of the restoration error. Define the error spectrum $E(\theta)$ by

$$E(\theta) = \mathcal{E} \left\{ \left| \sum_{k=0}^{m-1} \underline{e}_{k+1} \exp(-j\theta k) \right|^2 \right\}.$$

Then it follows that

$$E(\theta) = [1, \exp(j\theta), \ldots, \exp(j(m-1)\theta)]^T \mathbf{E} \begin{bmatrix} 1 \\ \exp(-j\theta) \\ \vdots \\ \exp(-j(m-1)\theta) \end{bmatrix},$$

and with

$$U_i(z) = \sum_{k=0}^{m-1} (\mathbf{u}_i)_{k+1} z^{-k}, \quad i = 1, \ldots, m,$$

one obtains

$$E(\theta) = (1 - \alpha)\sigma_n^2 \sum_{i=1}^{m} \frac{(1 - \lambda_i)^2}{\lambda_i} |U_i(\exp(j\theta))|^2. \tag{5.4}$$

The weighting factors $(1 - \lambda_i)^2/\lambda_i$ in (5.4) are very close to 0 for small i, since then the λ_i approach 1, and are very high for i close to m since then the λ_i approach 0. From the discussion in Appendix A it follows that the Fourier transform $U_1(\exp(j\theta))$ concentrates its energy in the interval $(\alpha\pi, \pi]$ and that the Fourier transform $U_m(\exp(j\theta))$ concentrates its energy in the interval $[0, \alpha\pi)$. It is shown in [50,6] that $\lambda_m \ll \lambda_{m-1}$ and therefore the weighting factor of

$$|U_m(\exp(j\theta))|^2$$

in (5.4) is much larger than the other weighting factors. From this it can be concluded that the spectral energy of the restoration error is concentrated in the low frequencies. And the error spectrum is approximated by

$$E(\theta) \cong (1 - \alpha)\sigma_n^2 \frac{(1 - \lambda_m)^2}{\lambda_m} |U_m(\exp(j\theta))|^2. \tag{5.5}$$

A similar result, namely that the restoration error in this case is pulse-shaped, is derived in [6]. The following example illustrates this behaviour.

i	λ_i	$\frac{(1-\lambda_i)^2}{\lambda_i}$
1	0.99E+00	0.10E−03
2	0.89E+00	0.14E−01
3	0.44E+00	0.71E+00
4	0.70E−01	0.12E+02
5	0.38E−02	0.26E+03
6	0.90E−04	0.11E+05
7	0.10E−05	0.10E+07
8	0.47E−08	0.21E+09

Table 5.1: Eigenvalues and weighting factors for a signal with a relative bandwidth of 0.7.

Example 6 *Consider a burst of length $m = 8$ and a signal band-limited to $\alpha\pi$, with $\alpha = 0.7$. Assume that the signal is quantized with step-size $\Delta = 1$, then it can be assumed that the quantization noise is uniformly distributed over the interval $[-\frac{1}{2}, \frac{1}{2}]$ and the noise has a variance $\sigma_n^2 = \frac{1}{12}$. The eigenvalues of $\tilde{\mathbf{G}}$ and the weighting factors of (5.3) are given in Table 5.1. Note that approximately a fraction α of the eigenvalues is close to 0 and a fraction $1 - \alpha$ is close to 1. This is a consequence of the Szegö limit theorem [28,29], see also Section 2.4. It is seen from Table 5.1 that the weighting factor of*

$$|U_8(\exp(j\theta))|^2$$

is approximately 200 times larger than the next largest weighting factor. It can also be seen that the relative error in the syndrome in (5.2) can be amplified 2×10^8 times and that the total variance of the error caused by the additive quantization noise comes down to approximately

$$(1 - \alpha)\sigma_n^2 \frac{(1 - \lambda_8)^2}{\lambda_8} = 0.53 \times 10^7.$$

The spectral energy of the restoration error is concentrated in the low frequencies. This is illustrated by Figure 5.1, showing the modulus of the Fourier transform

$$|U_8(\exp(j\theta))|^2$$

of $\mathbf{u_8} = [0.025, 0.143, 0.374, 0.583, 0.583, 0.374, 0.143, 0.025]^T$.

5.4   Results

From Example 6 it has become evident that the restoration method for band-limited signals is not suitable for restoring bursts in relatively wide-band signals even if the bursts are of moderate sizes. In this section

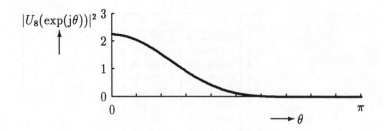

Figure 5.1: Modulus of the Fourier transform of the eigenvector associated with the smallest eigenvalue of $\mathbf{G}$, belonging to a signal with a relative bandwidth of 0.7.

the performance of this method is evaluated for small bursts of 4 and 6 samples and for a larger burst of 14 samples. The bandwidth is $\alpha\pi$, $\alpha = 0.54$. The following test signals were used in the evaluation:

1. **Multiple sinusoids.** A sequence of 512 samples, given by (3.36) in Section 3.8 was used. The patterns of the unknown samples were bursts of lengths $m = 4, 6, 14$.

2. **Digital audio signals.** Bursts of 4 and 6 unknown samples, occurring at a rate of 10 per second in a fragment of 36 seconds taken from a compact disc recording of Beethoven's Violin Concerto were restored. The sample frequency of the signal was 44.1 kHz.

3. **Prefiltered digital audio signals.** The signal and the patterns of unknown samples were the same as in test signal 2, apart from the fact that before the patterns of unknown samples were inserted, the signal was band-limited to half the sample frequency.

4. **Sinusoids corrupted by white noise.** Pseudo-random white noise was added to the sequences described under 3. Signal-to-noise ratios of 40 and 20 dB were considered. The patterns of the unknown samples were bursts of lengths $m = 4, 6, 14$.

These test signals are the same as in Chapter 3, so that a comparison between the methods can be made. For all test signals the performance is evaluated by means of the relative quadratic restoration error $\hat{E}$, defined in (3.37). As in Section 3.8, the results are presented in tables and figures. In the figures the original signals are marked by a (1), the restored signals are marked by a (2). Table 5.2 shows the performance on multiple sinusoids with various signal-to-noise ratios, test signals 1 and

	Signal-to-noise ratio					
	∞		40 dB		20 dB	
m	$\hat{E}$	E	$\hat{E}$	E	$\hat{E}$	E
4	0.53E−13	0	0.32E−01	0.46E−01	0.14E+01	0.46E+01
6	0.85E−06	0	0.16E+02	0.41E+01	0.21E+04	0.41E+03
14	0.86E−02	0	0.36E+03	0.50E+04	0.15E+05	0.50E+06

Table 5.2: Average restoration errors for various signal-to-noise ratios for a sum of sinusoids (test signals 1,4).

m	Non-prefiltered	Prefiltered
4	0.35E−01	0.10E−03
6	0.21E+01	0.88E−02

Table 5.3: Average restoration errors for digital audio signals, with and without prefiltering (test signals 2,3).

4. The relative quadratic restoration error $\hat{E}$ is an estimate for

$$E = (1 - \alpha)\frac{\sigma_n^2}{\sigma_s^2}\frac{(1 - \lambda_m)^2}{\lambda_m},$$

of which the value is included in Table 5.2. Table 5.3 shows the performance on the digital audio signal without prefiltering, test signal 2, and on the digital audio signal with prefiltering, test signal 3. Figure 5.2 shows a restoration result obtained on two sinusoids corrupted with white noise, test signal 4, the signal-to-noise ratio is 40 dB. Figure 5.3 shows a result obtained on the non-prefiltered digital audio signal, test signal 2. From the tables and figures it can be concluded that the restoration method for band-limited signals performs well if the number of unknown samples is low, say less than 5, and if it is guaranteed that the signal is band-limited. The method is very sensitive to out-of-band components in the signal. It can be observed from both figures that the error is pulse-shaped, as has been derived in [6] and Section 5.3.

An advantage of this method over the method for autoregressive processes of Chapter 3 is that if the bursts are small then fewer computations are required because the g_k, $k = -p, \ldots, p$, is known in advance and that for the allowed patterns of unknown samples the $\tilde{G}^{-1}$ can be computed and stored in advance. The total number of operations for one restoration then equals $O((2p + 1)m)$.

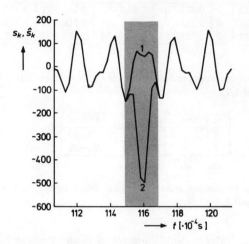

Figure 5.2: Original (1) and restored (2) segment of a sum of two sinusoids, signal-to-noise ratio=40 dB, $m = 6$, test signal 4.

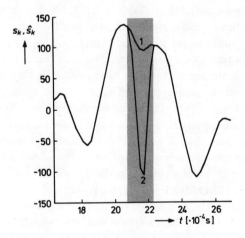

Figure 5.3: Original (1) and restored (2) segment of a digital audio signal without prefiltering, $m = 6$, test signal 2.

Chapter 6

Multiple sinusoids

6.1 Introduction

In this chapter a sample restoration method for signals consisting of r sinusoids is presented.[1] The restoration method is derived in Section 6.2. It is an example of a restoration method for signals that have a singular autocorrelation matrix. The method is iterative, like the restoration method for autoregressive processes, and has a similar performance. In the noise-free case it has a faster convergence, but is more sensitive to the presence of noise and to wrong assumptions about the number of sinusoids present in the signal. Results showing this behaviour are presented in Section 6.3. The method is not described and analyzed in the same detail as the methods presented in Chapters 3–5 because it has only been derived as a special case of the general restoration method of Chapter 2, to demonstrate that starting from a given signal model a restoration method can be found. The method has no obvious application, it is numerically complex, since it requires the computation of eigenvectors, and it seems more sensible to use in its place the method for autoregressive processes, which is computationally more attractive and less sensitive to noise.

6.2 The restoration method

First, linear minimum variance estimates for the unknown samples are derived, assuming that the signal spectrum or the autocorrelation function is known. In fact, in this case it is only required that the number of sinusoids present in the signal and their frequencies be known. After-

---

[1]A brief version of this chapter has been published previously in [11].

wards the method is made adaptive.

Assume that a segment of data s_k, $k = 1, \ldots, N$, is available, of which the samples at positions $t(1), \ldots, t(m)$ are unknown. This segment is a realization of a stationary stochastic process

$$\underline{s}_k = \sum_{i=1}^{r} A_i \sin(\theta_i k + \underline{\phi}_i), \quad k = -\infty, \ldots, +\infty.$$

The phases $\underline{\phi}_i$ have a uniform probability density function on the interval $[-\pi, \pi)$. The amplitudes A_i may also be stochastic, this is of no influence on the further results. The frequencies θ_i are in the interval $(0, \pi)$. The autocorrelation function of this signal is given by

$$R(k) = \sum_{i=1}^{r} \frac{A_i^2}{2} \cos(\theta_i k), \quad k = -\infty, \ldots, +\infty,$$

and the spectrum by

$$S(\theta) = \pi \sum_{i=1}^{r} \frac{A_i^2}{2} \left(\delta(\theta_i - \theta) + (\delta(\theta_i + \theta)) \right), \quad -\pi \leq \theta \leq \pi.$$

Figure A.2 in Appendix A shows an example of a spectrum of a signal containing two sinusoids. The $(p + 1) \times (p + 1)$ autocorrelation matrix $\mathbf{R}$, $p \geq 2r$, of this signal has maximum rank $2r$. The eigenvalues $\lambda_{2r+1}, \ldots, \lambda_{p+1}$ are all equal to 0. Furthermore, if $\mathbf{u}_{2r+1}, \ldots, \mathbf{u}_{p+1}$ are the eigenvectors associated with these eigenvalues then, with

$$U_i(z) = \sum_{k=0}^{p} (\mathbf{u}_i)_{k+1} z^{-k}, \quad i = 1, \ldots, p + 1,$$

one obtains

$$\begin{aligned} \lambda_i &= \mathbf{u}_i^T \mathbf{R} \mathbf{u}_i \\ &= \frac{1}{2\pi} \int_{-\pi}^{\pi} S(\theta) \left| U_i(\exp(j\theta)) \right|^2 d\theta \\ &= 0, \quad i = 2r + 1, \ldots, p + 1. \end{aligned} \qquad (6.1)$$

This shows that the Fourier transforms $U_i(\exp(j\theta))$, $i = 2r + 1, \ldots, p + 1$, have zeros at the frequencies of the sinusoids present in the signal, and also that with

$$G(\exp(j\theta)) = \left| U_i(\exp(j\theta)) \right|^2, \quad i \in \{2r + 1, \ldots, p + 1\},$$

or

$$g_k = \sum_{j=1}^{p+1-|k|} (\mathbf{u}_i)_j (\mathbf{u}_i)_{j+|k|}, \quad i \in \{2r + 1, \ldots, p + 1\}, \quad k = -p, \ldots, p,$$

a finite length g_k, $k = -p, \ldots, p$, is obtained. The g_k, $k = -p, \ldots, p$, obtained in this way can be used to build the system (2.13)

$$\tilde{G}\hat{x} = -z,$$

that can be solved for the unknown samples. Because $G(\exp(j\theta)) \geq 0$, $\tilde{G}$ is positive definite. If the number of sinusoids is not precisely known in advance, p must be chosen such that it is guaranteed that $p \geq 2r$ and

$$g_k = \sum_{j=1}^{p+1-|k|} (\mathbf{u}_{p+1})_j (\mathbf{u}_{p+1})_{j+|k|}, \quad k = -p, \ldots, p. \quad (6.2)$$

It is seen in Section 6.3 that in the presence of noise, too high a choice of p leads to poor results.

In general the eigenvectors, or even the $(p+1) \times (p+1)$ autocorrelation matrix $\mathbf{R}$, are unknown. In that case both $\mathbf{u}_{p+1}$ and $\mathbf{x}$ have to be estimated from the data. This can be done by using an adaptive version of this restoration method that consists in minimizing

$$Q(\lambda, \mathbf{u}, \mathbf{x}) = \sum_{k=p+1}^{N} \left(\sum_{l=0}^{p} u_{l+1} s_{k-l} \right)^2 - \lambda \left(\mathbf{u}^T \mathbf{u} - 1 \right) \quad (6.3)$$

as a function of λ, $\mathbf{u}$ and $\mathbf{x}$. As in Chapter 3, $x_i = s_{t(i)}$, $i = 1, \ldots, m$. The particular choice for minimizing $Q(\lambda, \mathbf{u}, \mathbf{x})$ to obtain estimates for $\mathbf{u}$ and $\mathbf{x}$ can be justified as follows. First, assuming that the data are complete, $Q(\lambda, \mathbf{u}, \mathbf{x})$ is written as

$$Q(\lambda, \mathbf{u}, \mathbf{x}) = \mathbf{u}^T \mathbf{C} \mathbf{u} - \lambda \left(\mathbf{u}^T \mathbf{u} - 1 \right). \quad (6.4)$$

Here

$$c_{i,j} = \sum_{k=p+1}^{N} s_{k-i+1} s_{k-j+1}.$$

Note that the definition of $c_{i,j}$ differs slightly from the one in (3.18) and that $\mathbf{C}$ is non-negative definite. This function has stationary points at $\mathbf{u} = \hat{\mathbf{u}}$, satisfying

$$\mathbf{C}\hat{\mathbf{u}} = \lambda \hat{\mathbf{u}}, \quad \mathbf{u}^T \mathbf{u} = 1. \quad (6.5)$$

From (6.5) it follows that $\hat{\mathbf{u}}$ is an eigenvector of $\mathbf{C}$. At the stationary points one has

$$Q(\lambda, \hat{\mathbf{u}}, \mathbf{x}) = \lambda. \quad (6.6)$$

From this it follows that $\hat{\mathbf{u}}$ must be chosen as the eigenvector associated with the minimum eigenvalue λ_{p+1} of $\mathbf{C}$. The matrix $\mathbf{C}$ is an unbiased estimate for $(N - p)\mathbf{R}$, and the minimization of $Q(\lambda, \mathbf{u}, \mathbf{x})$ as a function of $\mathbf{u}$ can be regarded as the estimation of $\mathbf{u}_{p+1}$ in (6.2).

Assuming that λ and $\mathbf{u}$ are known, $Q(\lambda, \mathbf{u}, \mathbf{x})$ is written as

$$Q(\lambda, \mathbf{u}, \mathbf{x}) = \sum_{k=p+1}^{N} \left(\sum_{l=0}^{p} u_{l+1} v_{k-l} \right)^2 + 2\mathbf{x}^T \mathbf{z} + \mathbf{x}^T \tilde{\mathbf{G}} \mathbf{x}. \qquad (6.7)$$

This function is the same as (3.16). It is minimized by choosing $\mathbf{x} = \hat{\mathbf{x}}$, where $\hat{\mathbf{x}}$ the solution of (2.13), which is the linear minimum variance estimate for $\mathbf{x}$ for known $\mathbf{u}$.

Minimizing (6.3) as a function of λ, $\mathbf{u}$ and $\mathbf{x}$ is even more complicated than minimizing (3.13), and in this case too an iterative procedure is used. Starting from an initial estimate $\hat{\mathbf{x}}^{(0)}$ for the unknown samples, $\hat{\mathbf{x}}^{(0)} = \mathbf{0}$ for instance, one computes subsequently in every iteration step, assuming that $\mathbf{x}$ is known, the minimum eigenvector of $\mathbf{C}$ and the eigenvector associated with it and, assuming that λ and $\mathbf{u}$ are known, an estimate for the unknown samples. As is the case with $Q(\mathbf{a}, \mathbf{x})$ in Chapter 3, $Q(\lambda, \mathbf{u}, \mathbf{x})$ decreases to some non-negative number, but it is difficult to check whether the minimum thus reached is global. It is shown in Section 6.3 that the number of iterations required for a good restoration result is usually 3.

In fact, $Q(\lambda, \mathbf{u}, \mathbf{x})$ should decrease to 0, but if noise is present in the signal, this value is not reached. If white noise with variance σ^2 is present in the signal, the matrix $\mathbf{C}$ is altered such that one has

$$\mathcal{E}\{\mathbf{C}\} = (N - p)(\mathbf{R} + \sigma^2 \mathbf{I}).$$

From this it follows that $\mathbf{u}$ remains unchanged and $\lambda \cong (N - p)\sigma^2$, which would seem to indicate that this method is rather insensitive to the presence of white noise. The results of Section 6.3 demonstrate that this is not true, especially if the number of sinusoids is assumed to be high. The reason for this is that the matrix that is added to $\mathbf{C}$ is not actually an identity matrix and its off-diagonal components distort the $\mathbf{u}$ [51].

Regarding efficient computation, this method is not attractive since in each iteration step an eigenvalue and an eigenvector have to be computed, which requires $O((p+1)^4)$ operations. A possibly more efficient iterative method is given in [52], but even then the number of operations required makes this method unattractive.

m	p	N	SNR	1 iteration	3 iterations
16	4	64	∞	0.19E−02	0.16E−27
16	10	64	∞	0.13E−03	0.15E−30
16	4	512	∞	0.65E−08	0.61E−28
16	10	512	∞	0.15E−07	
16	4	64	40 dB	0.58E−02	0.48E−02
16	10	64	40 dB	0.40E−01	0.12E−01
16	4	512	40 dB	0.94E−02	0.92E−02
16	10	512	40 dB	0.23E−02	0.12E−02
16	4	64	20 dB	0.76E+00	0.35E+00
16	10	64	20 dB	0.19E+01	0.66E+00
16	4	512	20 dB	0.96E+00	0.90E+00
16	10	512	20 dB	0.46E+00	0.18E+00

Table 6.1: Average restoration errors for various signal-to-noise ratios (test signals 1,2). For $p = 10$, $N = 512$ after 3 iterations in the noiseless case, the autocovariance matrix was almost singular.

6.3   Results

The following test signals were used to evaluate the performance of the restoration method for multiple sinusoids:

1. **Multiple sinusoids.** Sequences of 64 and 512 samples, given by (3.36) in Section 3.8 were used. The patterns of the unknown samples were bursts of lengths $m = 16$.

2. **Sinusoids corrupted by white noise.** Pseudo-random white noise was added to the sequences described under 3. Signal-to-noise ratios of 40 and 20 dB were considered. The patterns of the unknown samples were bursts of lengths $m = 16$.

These signals were restored using the method described in this chapter. The averaged relative quadratic restoration errors $\hat{E}$, defined in (3.37), after 1 and 3 iterations are presented in Table 6.1. Assuming the presence of 2 and 5 sinusoids, p is taken to be 4 and 10, respectively. From Table 6.1 it can be seen that in the noiseless case the method performs well; compared with the results of the restoration method for autoregressive processes shown in Tables 3.6 and 3.7 it shows a somewhat faster convergence. If noise is present and the number of sinusoids is assumed to be high, its performance is poorer than that of the method for autoregressive processes shown in Table 3.11.

Taking into account the complexity, the sensitivity to noise and the fact that even in the noiseless case the performance of the restoration method for multiple sinusoids is only slightly better than that of the restoration method for autoregressive processes, it can be concluded that the latter method is to be preferred to the former.

Chapter 7

Digital images

7.1 Introduction

In this chapter a sample restoration method is presented for monochrome digital images, a digital image being defined here as a two-dimensional sequence of samples $s_{i,j}$ called picture elements or *pixels*. According to the CCIR 601 standard, an image consists of 576 lines, each containing 720 pixels. The pixels are 8-bit integers in the range 0–255, representing luminance values. The luminance value 0 corresponds to a black pixel, the luminance value 255 corresponds to a white pixel. In $s_{i,j}$ the subscript i is the number of the line, the subscript j is the number of the pixel on that line. The $s_{i,j}$ can be seen as the elements of a 576×720 matrix.

The restoration method of this chapter is closely related to the one for autoregressive processes described in Chapter 3. As is the case with the method of Chapter 3, it assumes a predictive signal model and exploits the information contained in the correct pixels neighbouring the pixels to be restored. The method is capable of restoring single unknown pixels as well as random groups. Experiments have shown that it is possible to restore a block of 8×8 lost pixels embedded in a block of 20×20. In modern picture coding systems [12] blocks of 8×8 pixels are often coded and transmitted as a whole. In the event of an uncorrectable transmission error a complete block might be lost. An interesting application of the technique described here is concealment of these transmission errors. Another application is restoration of old, scratched pictorial material, such as films and photographs.

The method presented in this chapter is based on a predictive image model. This model is not merely a two-dimensional extension of the autoregressive process that is used as a model in Chapter 3, but it is adapted to the non-stationary character of the image signal. Section

7.2 describes the signal model used here. Parameter estimation for this model is discussed in Section 7.3.

The derivation of linear minimum variance estimates for the unknown samples, which amounts in this case to the derivation of a two-dimensional function $G(\theta_1, \theta_2)$, is omitted. Starting from the signal model of Section 7.2 the adaptive restoration method is presented directly. The restoration method is iterative and in each iteration step model parameters and pixel values are estimated. The restoration is such that the estimated pixel values fit the previously estimated model parameters as well as possible. Section 7.4 describes how this is done.

For evaluations, images from the popular sequences CAR and $TEENY$ have been used. Pictures and a description of the results are given in Section 7.5.

7.2   A predictive image model

The essence of linear prediction on images is that a pixel can be accurately estimated as a linear combination of neighbouring pixels. This is expressed mathematically by the following expression:

$$\sum_{\text{range}(k,l)} a_{k,l} s_{i-k,j-l} = e_{i,j}, \ a_{0,0} = 1. \tag{7.1}$$

As in Chapter 3, the coefficients $a_{k,l}$ are called *prediction coefficients*, $e_{i,j}$ is called the *prediction error*, and range(k, l) denotes the extent of the prediction filter. Since $a_{0,0} = 1$, (7.1) can be rewritten in such a way that $s_{i,j}$ is expressed as the sum of neighbouring pixels and $e_{i,j}$, the prediction error. The prediction coefficients are chosen such that the energy of the residual signal is minimal. The prediction is accurate if the prediction error is small compared with the pixel values. Predictive models are used in image coding systems [12,53].

In Chapter 3 it is assumed that the prediction error is white noise. In that case the signal is called an autoregressive process. If $e_{i,j}$ in (7.1) is zero-mean white noise the signal $s_{i,j}$ is a two-dimensional autoregressive process. However, due to the character of image signals, their prediction error is not white noise and images cannot be modelled as autoregressive processes. For the extent of the prediction filter, denoted by range(k, l), there are several possibilities [54], seven of which are important here. In the first case range(k, l) is defined by

$$0 \leq k \leq p, \ 0 \leq l \leq q. \tag{7.2}$$

This is called *upper-left quarter-plane prediction*. Here, p and q are positive integral numbers determining the number of prediction coefficients. In all four cases of quarter-plane prediction they are also the *vertical and horizontal orders of prediction*. In the second case range(k, l) is defined by

$$0 \leq k \leq p, \ -q \leq l \leq 0. \tag{7.3}$$

This is called *upper-right quarter-plane prediction*. In the third case range(k, l) is defined by

$$-p \leq k \leq 0, \ -q \leq l \leq 0. \tag{7.4}$$

This is called *lower-right quarter-plane prediction*. In the fourth case range(k, l) is defined by

$$-p \leq k \leq 0, \ 0 \leq l \leq q. \tag{7.5}$$

This is called *lower-left quarter-plane prediction*. The first four possibilities were the four types of quarter-plane prediction. In the fifth case range(k, l) is defined by

$$0 \leq k \leq p, \ -q \leq l \leq q, \ (k, l) \neq (0, -q), \ldots, (0, 1). \tag{7.6}$$

This is called *upper half-plane prediction*. The vertical order of prediction is now p, the horizontal order of prediction is $2q$. More precisely, this type of prediction is an example of *asymmetrical upper half-plane prediction*, because more information from the left is used than from the right. In the sixth case range(k, l) is given by

$$-p \leq k \leq p, \ 0 \leq l \leq q, \ (k, l) \neq (-p, 0), \ldots, (1, 0). \tag{7.7}$$

This is called *left-hand half-plane prediction*. The vertical order of prediction is now $2p$, the horizontal order of prediction is q. In the seventh case range(k, l) is given by

$$-p \leq k \leq p, \ -q \leq l \leq q. \tag{7.8}$$

This is called *four-sided prediction*. The vertical order of prediction now is $2p$, the horizontal order of prediction is $2q$. There are other possibilities [54], but they are not considered here.

In Figure 7.1 the extents of prediction of three of the seven cases mentioned above are depicted for $p = 2, q = 2$. Quarter-plane and half-plane prediction are used in image coding [12,53]. An extensive analysis of the consequences of the choice of range(k, l) is given in [54].

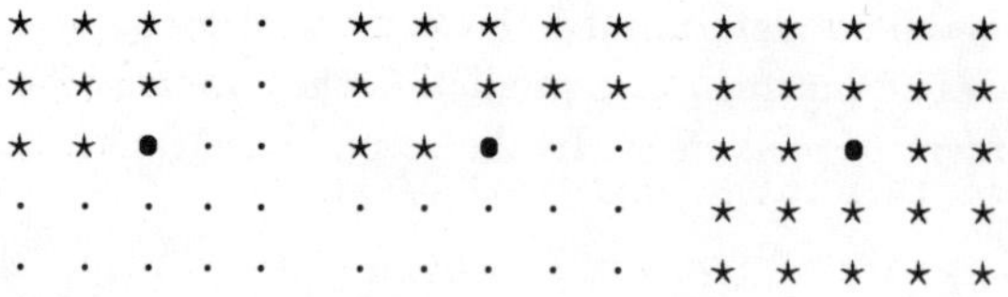

Figure 7.1: The extents of prediction of (from left to right) upper-left quarter-plane prediction, upper half-plane prediction and four-sided prediction. The predicted pixel is denoted by a • and the pixels used to predict this pixel are denoted by a ★. The other pixels are denoted by a ·.

The predictive model described so far is not well-suited to describe digital images accurately. In the first place, an image is not a stationary signal. Its properties are not constant over the image. The set of prediction coefficients is only locally valid, say for a block of up to 20×20 pixels. This means that only for that block does it produce a prediction error with minimal energy. For the next block a new set of coefficients is required. In the second place, the pixel values are luminance values. They are usually represented in the range $[0, 255]$. Their mean value is not equal to zero and this deviation from the model influences the quality of prediction. Also the mean value is not constant over an image, but varies spatially. A better predictive model is given by

$$\sum_{\text{range}(k,l)} a_{k,l}(s_{i-k,j-l} - \mu) = e_{i,j}, \tag{7.9}$$

$$a_{0,0} = 1,$$
$$1 \le i \le M,$$
$$1 \le j \le N.$$

Here M and N are the dimensions of the block for which the predictive model is valid and μ denotes the mean value within the block. For small blocks, say up to 8×8 pixels, μ can be considered a constant; for larger blocks the assumption that the pixel values are fluctuating around a mean value is not always true. In that case it can be assumed that the pixel values fluctuate around a slanted plane. The model is then given by

$$\sum_{\text{range}(k,l)} a_{k,l}(s_{i-k,j-l} - (k_1 + k_2(i - k) + k_3(j - l))) = e_{i,j}, \tag{7.10}$$

$$a_{0,0} = 1,$$
$$1 \le i \le M,$$
$$1 \le j \le N.$$

In (7.10) the plane is described by the expression $k_1 + k_2 i + k_3 j$ and k_1, k_2, k_3 are called the *plane coefficients*.

7.3   Parameter estimation

In this section methods used to estimate the parameters introduced in Section 7.2 are discussed. Throughout the section it will be assumed that there are no unknown samples. The methods discussed in this section are of use in Section 7.4, where the restoration method is made adaptive and parameters as well as unknown pixels have to be estimated.

The parameters of the models (7.1), (7.9) and (7.10) are the orders of prediction, or the related integers p, q, the prediction coefficients $a_{i,j}$, and the mean value μ or the plane coefficients k_1, k_2, k_3. The accuracy of the estimates for these parameters is determined by the estimation method and by the dimensions M, N of the block of data. For a stationary image a larger block means an increased accuracy of estimated parameters. Unfortunately, realistic and interesting images are seldom stationary and if the block's dimensions exceed certain bounds the prediction error will increase. The reason is that the variety of structures that can be present in the block cannot be modelled with one set of prediction coefficients. The values of M, N for which a block can be considered stationary depend on the kind of image and no strict rules can be given as to how to choose M, N. In this book the maximum values for M, N that are used are $M = 20, N = 20$, but it is questionable if a block of this size can still be considered stationary.

Another problem is the estimation of the orders of prediction. The best-known method is described in [38], but its practicality is questionable. In Chapter 3 the order of prediction is a function of the number of unknown samples; in the present case good results are obtained with fixed orders of prediction, determined by the choice $p = 2, q = 2$.

In the case of the simplest model, given in (7.1), the prediction coefficients can be obtained by minimizing

$$Q_{\mathbf{a}}(\mathbf{a}) = \sum_{\text{range}(i,j)} \left(\sum_{\text{range}(k,l)} a_{k,l} s_{i-k,j-l} \right)^2 = \sum_{\text{range}(i,j)} e_{i,j}^2. \qquad (7.11)$$

Here $\mathbf{a}$ is a vector in which the unknown prediction coefficients have been arranged, see Appendix G, and range(i, j) denotes those $e_{i,j}$ that can be calculated given a block of $M \times N$ pixels and range(k, l) in (7.1). This expression is quadratic in the unknown prediction coefficients and can be minimized by taking the derivatives with respect to the unknowns, setting these equal to zero and solving the estimates $\hat{\mathbf{a}}$ for the prediction coefficients from the resulting linear system of equations. This procedure for estimating prediction coefficients is the same as the one of Chapter 3. It was mentioned there that efficient system-solving algorithms exist for the latter step. In the present case that is still an open question. In Chapter 3 the system matrix is the signal's autocovariance matrix, which structure can be exploited when solving the system. In the present case, although this matrix also contains autocovariance coefficients, its structure is less obvious and it is not clear if it can be exploited similarly. In [53,54] other methods are suggested for estimating prediction coefficients for two-dimensional prediction, some of which can be implemented efficiently. More details on the system of equations that has to be solved are given in Appendix G. The minimum $Q_{\mathbf{a}}(\hat{\mathbf{a}})$ of $Q_{\mathbf{a}}(\mathbf{a})$, the *estimated prediction error energy*, is often used as an indication of the quality of the prediction. In that case, the lower $Q_{\mathbf{a}}(\hat{\mathbf{a}})$, the better one considers the quality of prediction.

Minimizing the energy of the prediction error cannot be used to estimate $a_{k,l}$ and μ or k_1, k_2, k_3 in the models (7.9) or (7.10), because in these cases this function is not quadratic in the unknowns. An iterative minimization procedure can be tried, but this would lead to an undesirable increase of complexity. The following procedures give good results. In the case of the model (7.9), μ is estimated by minimizing

$$Q_\mu(\mu) = \sum_{i=1}^{M} \sum_{j=1}^{N} (s_{i,j} - \mu)^2. \tag{7.12}$$

The result is of course that the estimate $\hat{\mu}$ for μ is the mean value of the block. In the case of the model (7.10) $\mathbf{k} = [k_1, k_2, k_3]^T$ is estimated by minimizing

$$Q_{\mathbf{k}}(\mathbf{k}) = \sum_{i=1}^{M} \sum_{j=1}^{N} (s_{i,j} - (k_1 + k_2 i + k_3 j))^2. \tag{7.13}$$

As a result the estimate: $\hat{\mathbf{k}} = [\hat{k}_1, \hat{k}_2, \hat{k}_3]^T$ for $\mathbf{k}$ can be solved from the following set of linear equations

$$\begin{bmatrix} \sum_{i,j} & \sum_{i,j} i & \sum_{i,j} j \\ \sum_{i,j} i & \sum_{i,j} i^2 & \sum_{i,j} ij \\ \sum_{i,j} j & \sum_{i,j} ij & \sum_{i,j} j^2 \end{bmatrix} \begin{bmatrix} \hat{k}_1 \\ \hat{k}_2 \\ \hat{k}_3 \end{bmatrix} = \begin{bmatrix} \sum_{i,j} s_{i,j} \\ \sum_{i,j} i s_{i,j} \\ \sum_{i,j} j s_{i,j} \end{bmatrix}. \tag{7.14}$$

The summations in (7.14) are over all the known pixels in the block. After computation of $\hat{\mu}$ or $\hat{k}_1, \hat{k}_2, \hat{k}_3$ the *auxiliary signal*

$$\tilde{s}_{i,j} = s_{i,j} - \hat{\mu}, \tag{7.15}$$

in the case of (7.9) and

$$\tilde{s}_{i,j} = s_{i,j} - (\hat{k}_1 + \hat{k}_2 i + \hat{k}_3 j) \qquad (7.16)$$

in the case of (7.10) can be substituted for $s_{i,j}$ in (7.9) or (7.10), respectively. Then the same approach as in the case of (7.1) can be taken and (7.11), with $\tilde{s}_{i,j}$ substituted for $s_{i,j}$, can be minimized as a function of the prediction coefficients.

7.4 The restoration method

It is assumed that a block containing the pixels $s_{i,j}$, $i = 1, \ldots, M$, $j = 1, \ldots, N$, is available. In this block m pixels at the positions $(t_1(i), t_2(i))$, $i = 1, \ldots, m$, are unknown. To ensure that the unknown pixels are embedded in a sufficiently large neighbourhood of known pixels it is required that for $i = 1, \ldots, m$

$$p < t_1(i) < M - p + 1, \ q < t_2(i) < N - q + 1, \qquad (7.17)$$

in the cases of quarter-plane prediction,

$$p < t_1(i) < M - p + 1, \ 2q < t_2(i) < N - 2q + 1, \qquad (7.18)$$

in the case of upper half-plane prediction,

$$2p < t_1(i) < M - 2p + 1, \ q < t_2(i) < N - q + 1, \qquad (7.19)$$

in the case of left-hand half-plane prediction and

$$2p < t_1(i) < M - 2p + 1, \ 2q < t_2(i) < N - 2q + 1, \qquad (7.20)$$

in the case of four-sided prediction. If the conditions in (7.17), (7.18), (7.19) or (7.20) are not satisfied, good restoration is still possible, but the results are notationally less satisfactory. For notational convenience the unknown pixels are in a vector $\mathbf{x}$, defined by

$$x_i = s_{t_1(i),t_2(i)}, \ i = 1, \ldots, m. \qquad (7.21)$$

The estimate for $\mathbf{x}$ is denoted by $\hat{\mathbf{x}}$.

First it is assumed that the model parameters are known. As a model the predictive model of (7.10) is used. Of course, the other two models (7.1) and (7.9) can also be used, but the derivations are similar and a separate discussion will not provide more insight. Furthermore, the model of (7.10) is the most elaborate and has given the best looking results. In general, the model parameters are not known in advance, but have to be determined from the available, but incomplete, data. Therefore a method is given for determining both the parameters and the lost pixels from the available data for some extent of prediction.

Estimates for the unknown pixels are chosen such that the resulting signal satisfies the assumed predictive model as well as possible in a quadratic sense. This means that, like the method of Chapter 3, the expression

$$Q_{\mathbf{x}}(\mathbf{x}) = \sum_{\text{range}(i,j)} (\sum_{\text{range}(k,l)} a_{k,l}(s_{i-k,j-l} - (k_1 + k_2(i-k) + k_3(j-l))))^2 \quad (7.22)$$

is minimized as a function of $\mathbf{x}$.

The function $Q_{\mathbf{x}}(\mathbf{x})$ is quadratic in the elements of $\mathbf{x}$. It can be minimized by setting the derivatives with respect to the elements of $\mathbf{x}$ equal to zero and by solving $\hat{\mathbf{x}}$ from the resulting system

$$\tilde{\mathbf{G}}\hat{\mathbf{x}} = -\mathbf{z}$$

of m equations with m unknowns. This system is denoted in the same way as the system (2.13).

The positions of the unknown samples satisfy one of the relations (7.17), (7.18), (7.19) or (7.20). Because of this, the system matrix and the right-hand side of the set of equations take a convenient form, as in (2.19). First define

$$g_{i,j} = \sum_{\text{range}(k,l)} a_{k,l} a_{k+i,l+j}. \quad (7.23)$$

Here range(k, l) and range(i, j) are chosen such that the $a_{k,l}$ and $a_{k+i,l+j}$ in the summation are defined. For range(i, j) it can be derived that

$$-p \leq i \leq p, -q \leq j \leq q, \quad (7.24)$$

in the cases of quarter-plane prediction,

$$-p \leq i \leq p, \ -2q \leq j \leq 2q, \quad (7.25)$$

in the case of upper half-plane prediction,

$$-2p \leq i \leq 2p, \ -q \leq j \leq q, \quad (7.26)$$

in the case of left-hand half-plane prediction, and

$$-2p \leq i \leq 2p, \ -2q \leq j \leq 2q, \quad (7.27)$$

in the case of four-sided prediction. Now $\tilde{\mathbf{G}}$ follows from

$$\tilde{\mathbf{G}}_{i,j} = g_{t_1(i)-t_1(j),t_2(i)-t_2(j)}, \quad i,j = 1,\ldots,m, \qquad (7.28)$$

and $\mathbf{z}$ follows from

$$z_i = \sum_{k,l} g_{k,l}\tilde{v}_{k-t_1(i),l-t_2(i)}. \qquad (7.29)$$

Here $\tilde{v}_{i,j}$ is the auxiliary signal $\tilde{s}_{i,j}$, with zeros at the positions of the unknown pixels. The range of k and l in this summation are such that the $g_{k,l}$ in the summation are defined. The vector $\mathbf{z}$ in (7.29) is called the *syndrome*.

Estimation of the unknown pixels will now be discussed for the case where the parameters are also unknown. The parameters involved are p, q, the plane coefficients k_1, k_2, k_3 and the prediction coefficients $a_{i,j}$. Next to these, the block's dimensions M, N must be chosen. The orders of prediction are chosen fixed, the other parameters are estimated. The block's dimensions are chosen in dependence on the number of unknown pixels. The number of known pixels should be such that reliable estimates for the prediction coefficients can be made. There is no strict rule as to how to determine this number, but good results are obtained for instance with a block of 20×20 pixels with 8×8 pixels missing and with a block of 16×16 pixels with 4×4 pixels missing. The choice of M, N is not very critical, but if they are chosen too large the block of pixels can no longer be considered stationary and the prediction coefficients will not correctly represent the local structure in the block. On the other hand, if they are chosen too small, the estimates for the prediction coefficients will not be accurate. The orders of prediction have been chosen according to $p = 2, q = 2$. This choice has given the best looking results. Smaller p and q give errors on the edges. Larger p and q give inaccurate estimates for the prediction coefficients, because the ratio of the amount of data to the number of prediction coefficients becomes unfavourable.

Unfortunately the parameters and the lost pixels cannot be estimated independently. As in Chapter 3, an iterative estimation procedure is tried with good results. Because the plane coefficients are also involved, the procedure described here differs slightly from the one of Chapter 3.

First, an initial estimate for $\mathbf{k}$, $\mathbf{k}^{(0)}$ is computed from the incomplete data. This is done by minimizing $Q_{\mathbf{k}}(\mathbf{k})$ in (7.13), leaving the terms containing unknown samples out of the summation. The next step might be to determine initial estimates for the unknown pixels by choosing

$$\hat{s}^{(0)}_{t_1(i),t_2(i)} = \hat{k}^{(0)}_1 + \hat{k}^{(0)}_2 t_1(i) + \hat{k}^{(0)}_3 t_2(i), \quad i = 1,\ldots,m, \qquad (7.30)$$

and then to derive first estimates for the prediction coefficients by applying the method described in Section 7.3. Better results, however, are obtained by choosing an initial estimate $\hat{\mathbf{a}}^{(0)}$ for the prediction coefficients, defined by

$$a_{0,0}^{(0)} = 1, \ a_{0,1}^{(0)} = -\alpha, \ a_{1,0}^{(0)} = -\alpha, \ a_{1,1}^{(0)} = \alpha^2, \qquad (7.31)$$
$$a_{i,j}^{(0)} = 0, \ \text{otherwise}.$$

Here α is a real number close to 1, for instance $\alpha = 0.999$. On substitution of $\hat{\mathbf{k}}^{(0)}$ and $\hat{\mathbf{a}}^{(0)}$ into $Q_{\mathbf{x}}(\mathbf{x})$ in (7.22) an initial estimate $\hat{\mathbf{x}}^{(0)}$ for the unknown pixels is derived by minimizing this function as has been described previously in this section. A possible explanation for the fact that initial estimates for the unknown pixels obtained in this way give better results than those from (7.30) is that the estimated prediction coefficients adapt to structures, say edges, in the block. If the initial estimates of (7.30) are used, artificial edges can be introduced at the boundaries of the error pattern. This would bias the estimated prediction coefficients thereby leading to a biased result. If the initial estimates for the prediction coefficients of (7.31) are used to obtain initial estimates for the unknown pixels, then the transitions in the luminance values at the boundaries of the error pattern are smooth, and no artificial edges are introduced. In fact, in many cases, for instance when the number of unknown pixels is small, or when there is not enough time to compute better estimates, or when a medium quality restoration is good enough, calculations can be stopped here and this initial estimate can be used as a final one.

Once an initial estimate for the unknown pixels has been obtained, the procedure is continued as follows. The estimated pixel values are substituted into $s_{i,j}$, and $Q_{\mathbf{k}}(\mathbf{k})$ is minimized as a function of $\mathbf{k}$ in the manner that has been described in Section 7.3. This leads to a first estimate $\hat{\mathbf{k}}^{(1)}$ for the $\mathbf{k}$. Then $\tilde{s}_{i,j}$, (7.16), is calculated and substituted into $Q_{\mathbf{a}}(\mathbf{a})$ in (7.11). This function is minimized as a function of $\mathbf{a}$ as has been described in Section 7.3. This leads to a first estimate $\hat{\mathbf{a}}^{(1)}$ for the prediction coefficients. These are substituted into $Q(\mathbf{x})$ in (7.22), which is minimized as a function of $\mathbf{a}$. This leads to a first estimate $\hat{\mathbf{x}}^{(1)}$ for the unknown pixels.

This procedure can be continued by calculating $\hat{\mathbf{k}}^{(2)}$, $\hat{\mathbf{a}}^{(2)}$, $\hat{\mathbf{x}}^{(2)}$, $\hat{\mathbf{k}}^{(3)}$, $\hat{\mathbf{a}}^{(3)}$, $\hat{\mathbf{x}}^{(3)}$ and so on, until a satisfactory result is obtained. In practice three iterations give good results.

The iterative restoration method can be used with one of the extents of prediction mentioned in Section 7.2. Experiments show that if a predictive model based on, for instance, upper half-plane prediction (7.6) is used, some errors may occur in blocks containing horizontal structures. On the other hand, if left-hand half-plane prediction, (7.7) is used, errors may occur in blocks containing vertical structures. In general, it can be said that the choice for one extent of prediction introduces errors in structures that cannot be described with a prediction filter of this extent.

At first, it was expected that the errors mentioned in the previous paragraph could be avoided by using four-sided prediction (7.8), but this turned out not to be true. Two other approaches have been tried. First, two initial estimates for the prediction coefficients were made, $\hat{a}_u^{(1)}$ and $\hat{a}_l^{(1)}$, assuming respectively an upper and a left-hand half-plane predictive model. The model with the lowest estimated prediction error energy was then chosen to be used subsequently. This gave a clearly visible improvement. However, the best results were obtained as follows. Four initial estimates for the prediction coefficients were made, $\hat{a}_{ul}^{(1)}$, $\hat{a}_{ur}^{(1)}$, $\hat{a}_{lr}^{(1)}$ and $\hat{a}_{ll}^{(1)}$, assuming respectively an upper-left, upper-right, lower-right and a lower-left predictive model. The model with the lowest estimated prediction error energy was then chosen for subsequent use. A possible explanation for the phenomenon that the simplest models perform best is that the ratio of parameters to available data is higher than in the case of half-plane or four-sided prediction, which leads to better estimates for the prediction coefficients.

The method that has been presented in this chapter can be summarized as follows.

1. Choose the block sizes M, N, depending on the error pattern. For instance, $M = 20, N = 20$ for a block of 8×8 lost pixels and $M = 16, N = 16$ for a block of 4×4 missing pixels.

2. Choose $p = 2, q = 2$.

3. Compute initial estimates $\hat{k}^{(0)}$, $\hat{a}^{(0)}$ and $\hat{x}^{(0)}$.

4. Compute the quarter-plane predictors $\hat{a}_{ul}^{(1)}$, $\hat{a}_{ur}^{(1)}$, $\hat{a}_{lr}^{(1)}$ and $\hat{a}_{ll}^{(1)}$ and their respective total prediction error energies $Q_a(\hat{a}_{ul}^{(1)})$, $Q_a(\hat{a}_{ur}^{(1)})$, $Q_a(\hat{a}_{lr}^{(1)})$ and $Q_a(\hat{a}_{ll}^{(1)})$. Choose for further estimation the quarter-plane with the smallest prediction error energy.

5. Compute iteratively $\hat{k}^{(2)}, \hat{a}^{(2)}, \hat{x}^{(2)}, \hat{k}^{(3)}, \hat{a}^{(3)}, \hat{x}^{(3)}$ etc. until a satisfactory result is obtained. In practice three iterations give good results.

Whether or not this method has any practical value is determined by its numerical properties. The *computational complexity*, or number of operations required for one restoration, must be known as well as the *numerical precision* required for each operation. If the computational complexity is high, it is of interest to know whether efficient algorithms exist for the steps into which the method can be divided.

As has been done for the method of Section 3 [41], the numerical precision can be assessed experimentally. This has still to be done for the present method.

From now on only quarter-plane prediction is considered. To build the complete system (7.14) from which the plane coefficients are solved, $3MN$ operations are required. Note that the matrix elements, in fact even the inverse of the system matrix, can be stored in advance and need not be computed. This is not true when an initial estimate for the plane coefficients is calculated because then the unknown pixels, which may have varying positions, are not taken into account in the summations. The number of additional operations that is required for this is small and will be omitted. The most interesting application of the algorithm is restoration of a block of 8 × 8 pixels to conceal transmission errors in an image coding system. In that case the system matrix is also known in advance because there is a fixed error pattern.

In Appendix G it is shown how to build the system of equations from which the prediction coefficients are solved. From (G.6) in Appendix G it can be seen that the number of operations required to build this system equals $r(r+1)(M-p)(N-q)$, where r is the number of prediction coefficients. It can be shown that this number can be reduced to approximately

$$r\min(p+1, q+1)(M-p)(N-q).$$

To build the system (2.13) the coefficients $g_{i,j}$ in (7.23) have to be calculated. This requires

$$\frac{1}{2}(p+1)^2(q+1)^2 - \frac{1}{2}(p+1)(q+1)$$

operations. The syndrome (7.29) is the result of a convolution of the coefficients $g_{i,j}$ with the known pixels, which requires $m(2p+1)(2q+1)$ operations.

There are three system-solving operations: *calculation of the plane coefficients*, *calculation of the prediction coefficients* and *calculation of the lost pixels*. In the calculation of the plane coefficients the inverse of the system matrix can be stored in advance and only nine operations are required.

In all other cases[1] the system matrix is positive definite and the systems can be solved by Cholesky decomposition, whose numerical prop-

---

[1]In fact, including the case of the calculation of the plane coefficients.

erties have been studied extensively [22]. The number of operations required for this type of system-solving is $O(\frac{1}{6}n^3)$, where n is the number of equations and unknowns.

The number of operations required for calculating the prediction coefficients using Cholesky decomposition is approximately $O(\frac{1}{6}r^3)$, which is relatively small since there are only eight prediction coefficients. The calculation of prediction coefficients in this manner is generally referred to as the *autocovariance method* [53]. Efficient methods of solving the system exist for the one-dimensional case, but it is not clear whether these methods also exist for the two-dimensional case. If in (G.6) the pixel values outside the block of data are set equal to zero and if the summation is over all the integers, the method is called the *autocorrelation method*. For this method it can be shown that the system matrix is block-Toeplitz and efficient system-solving algorithms using a number of operations $O(\min(p+1, q+1)r^2)$ exist [47]. For small p, q it may very well be that Cholesky decomposition is the most efficient method of solving the system.

The number of lost pixels may be as high as 64, and in this case the number of calculations needed to solve the system for the estimates of the lost pixel values becomes very high. If the lost pixels occur only in blocks, then the systems matrix is block-Toeplitz and the number of operations required to solve the system is $O(m^{2\frac{1}{2}})$ [47].

If it is assumed that $M = N = 20$, $p = q = 2$ and that there is a block of 8 × 8 unknown pixels, then the number of operations required per iteration is approximately 44,000, to which the estimation of the lost pixels contributes approximately 33,000 operations. Also, it turns out that Cholesky decomposition is the most efficient way to obtain the prediction coefficients. Including the computation of the initial estimate, some 176,000 operation are required to restore a block of 8 × 8 pixels. For an image containing 576 lines of 720 pixels this means that concealment of one block increases the number of operations per pixel by ±0.43. If the maximum allowed increase is 5 operations per pixel, then 11 blocks, or 0.17% of the blocks can be restored.

7.5   Results

This section gives results of the final restoration method as presented in Section 7.4. The test images used here were an image from the sequence $TEENY^2$ and an image from the sequence CAR^3, both according to the

---

[2]Popular at Philips Research Laboratories.
[3]Often used in the European image processing community.

CCIR 601 standard. The original test images are shown in Figures 7.2 and 7.11. The results were obtained by computer simulations. Two types of error were restored: blocks of 4×4 unknown pixels and blocks of 8×8 unknown pixels. In the first case the images were divided into blocks of 16×16 pixels of which the centre blocks of 4×4 pixels were distorted. In the second case the images were divided into blocks of 20×20 pixels of which the centre blocks of 8×8 pixels were distorted. The distorted images are shown in Figures 7.3, 7.7, 7.12, 7.16. For both error patterns the initial restoration and the result after three iterations were computed, using block sizes of 16×16 pixels for the error patterns of 4×4 pixels and block sizes of 20×20 pixels for the error patterns of 8×8 pixels. For the orders of prediction $p = 2, q = 2$ is chosen. The restoration results are shown in Figures 7.4, 7.5, 7.8, 7.9, 7.13, 7.14, 7.17, 7.18. Figures 7.6, 7.10, 7.15, 7.19 show the differences (magnified 4 times) between the reconstructed images obtained after three iterations and the original images.

The results obtained after three iterations on the error patterns of 4×4 pixels can be considered very good; hardly any errors can be seen. The results obtained after three iterations on the error patterns of 8×8 pixels are less good; some errors can be seen, but they can still be qualified as reasonably good. The restoration of the image taken from the sequence CAR in this case shows more errors than the restoration of the image taken from the sequence $TEENY$. It can be observed that errors occur when the block of pixels that is used to determine the parameters contains a complicated structure, such as crossing lines or a digit from the car's number plate.

Looking at the differences between the restored and the original images, it can be observed that more errors occur than are directly seen. The method seems to generate restored pixels, which may not always be the right ones but which the human observer does not perceive as erroneous.

If the local structure of the image is not too complicated, and if very good results are not required, then the initial estimates can serve as restorations. It must be remarked that the results for error patterns of 8×8 pixels are rather poor, especially in the image taken from the sequence CAR.

Figure 7.2: Original image from sequence *TEENY*.

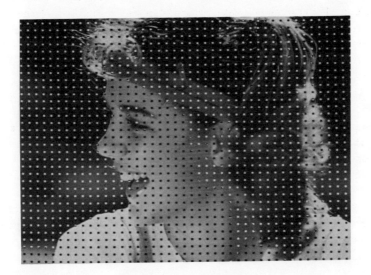

Figure 7.3: Distorted image from sequence *TEENY*. The error pattern contains 4 × 4 pixels.

Figure 7.4: Reconstructed image from sequence *TEENY*. The error pattern contains 4 × 4 pixels. The initial estimates are shown.

Figure 7.5: Reconstructed image from sequence *TEENY*. The error pattern contains 4 × 4 pixels. The results after three iterations are shown.

Figure 7.6: Reconstruction errors from sequence *TEENY*. The error pattern contains 4 × 4 pixels. The results after three iterations are shown.

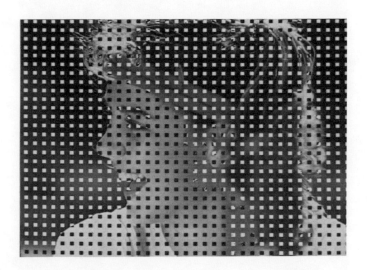

Figure 7.7: Distorted image from sequence *TEENY*. The error pattern contains 8 × 8 pixels.

Figure 7.8: Reconstructed image from sequence *TEENY*. The error pattern contains
8 × 8 pixels. The initial estimates are shown.

Figure 7.9: Reconstructed image from sequence *TEENY*. The error pattern contains
8 × 8 pixels. The results after three iterations are shown.

Figure 7.10: Reconstruction errors from sequence *TEENY*. The error pattern contains 8 × 8 pixels. The results after three iterations are shown.

Figure 7.11: Original image from sequence *CAR*.

Figure 7.12: Distorted image from sequence *CAR*. The error pattern contains 4×4 pixels.

Figure 7.13: Reconstructed image from sequence *CAR*. The error pattern contains 4×4 pixels. The initial estimates are shown.

Figure 7.14: Reconstructed image from sequence *CAR*. The error pattern contains 4 × 4 pixels. The results after three iterations are shown.

Figure 7.15: Reconstruction errors from sequence *CAR*. The error pattern contains 4 × 4 pixels. The results after three iterations are shown.

Figure 7.16: Distorted image from sequence *CAR*. The error pattern contains 8 × 8 pixels.

Figure 7.17: Reconstructed image from sequence *CAR*. The error pattern contains 8 × 8 pixels. The initial estimates are shown.

Figure 7.18: Reconstructed image from sequence *CAR*. The error pattern contains 8 × 8 pixels. The results after three iterations are shown.

Figure 7.19: Reconstruction errors from sequence *CAR*. The error pattern contains 8 × 8 pixels. The results after three iterations are shown.

Chapter 8

Concluding remarks

In this book methods are discussed for restoring unknown sample values with known positions in various kinds of signals. All these methods can be regarded as special cases of a general linear minimum variance estimation method. This general method is discussed separately, but it is of little practical value because it requires knowledge of the signal spectrum or the signal's autocorrelation function. It is to be seen as an analysis tool providing insight into the sample restoration problem; for instance, it supplies information on the restoration error.

Five adaptive restoration methods are discussed as special cases of this general method; they are sample restoration methods for autoregressive processes, speech signals, band-limited signals, sums of sinusoids and digital images. The restoration methods for autoregressive processes, speech signals and digital images are the most successful. The restoration method for autoregressive processes can also be used to restore successfully speech signals, band-limited signals and sums of sinusoids; moreover the method for sample restoration in digital images can be seen as an extension of this method. However, for speech signals, in which very large bursts of unknown samples can occur, its computation is rather involved. Therefore another method is developed especially for speech signals.

The method for sample restoration in band-limited signals is extremely sensitive to the presence of noise or other out-of-band components in the signal and is only suitable for the restoration of small patterns of unknown samples in narrow-band signals. Only in these cases does it give results that are better than the results obtained with the sample restoration method for autoregressive processes. An advantage then is that it requires far fewer computations than any of the other methods. The restoration method for multiple sinusoids is also rather sensitive to the presence of noise. Furthermore, a good knowledge of the number of sinusoids present in the signal is required. Compared with the restoration method for autoregressive processes, it gives slightly better results on noiseless signals, but if the signals are corrupted by white noise the restoration method for autoregressive processes turns out to be better. Apart from that, the restoration methods for sums of sinusoids requires the computation of eigenvalues and eigenvectors, thus making it unattractive.

The sample restoration method for unknown pixels in digital images can be seen as a two-dimensional extension of the restoration method for autoregressive processes, which is specially adapted to the non-stationary character of digital image signals.

The restoration methods for autoregressive processes, speech signals and digital images are designed with certain applications in mind. The method for autoregressive processes is designed to restore lost samples in digital audio signals, such as may occur in a compact disc signal; the method for speech signals can be used in mobile automatic telephony, and the method for digital images can be used to restore transmission errors in digital images that have been coded with a block-based transform coding method.

Discussion of the feasibility of a hardware implementation concentrates mainly on the two methods for restoring autoregressive processes and speech. Some implementions require more elaborate arithmetic, such as system-solving. This is more complicated than the now commonly used digital filtering operations, but it can be realized in state-of-the-art technology. Therefore, if the applications are sufficiently important, then the design engineers should consider implementing these methods.

Appendix A

The autocorrelation matrix

In this appendix the conditions under which the $N \times N$ autocorrelation matrix $\mathbf{R}$ is singular are investigated. First it must be remarked that $\mathbf{R}$ is *non-negative definite*, which means that for every $\mathbf{u} \in \mathbb{R}^N$

$$\mathbf{u}^T \mathbf{R} \mathbf{u} = \sum_{k,l=1}^{N} R(k - l) u_k u_l \geq 0. \tag{A.1}$$

Here $R(k)$, $k = -\infty, \ldots, +\infty$, is the signal's autocorrelation function. A proof is given in [23]. If only the $>$-sign holds in (A.1), $\mathbf{R}$ is called *positive definite*. Equivalent to the statement that $\mathbf{R}$ is non-negative definite for any N is the statement that the signal spectrum $S(\theta)$, $-\pi \leq \theta \leq \pi$, is non-negative. This is also shown in [21,23]. The autocorrelation matrix $\mathbf{R}$ is singular if for some $\mathbf{u} \in \mathbb{R}^N$, $\mathbf{u} \neq \mathbf{0}$

$$\mathbf{u}^T \mathbf{R} \mathbf{u} = 0.$$

Define $U(z)$ by

$$U(z) = \sum_{k=0}^{N-1} u_{k+1} z^{-k}, \quad z \in \mathbb{C},$$

then it can be shown that

$$\mathbf{u}^T \mathbf{R} \mathbf{u} = \frac{1}{2\pi} \int_{-\pi}^{\pi} S(\theta) \left| U(\exp(j\theta)) \right|^2 d\theta. \tag{A.2}$$

Since $S(\theta) \geq 0$ and $|U(\exp(j\theta))|^2 \geq 0$, for $-\pi \leq \theta \leq \pi$, $\mathbf{u}^T \mathbf{R} \mathbf{u}$ can only be equal to zero if $U(\exp(j\theta))$ is non-zero in the regions where $S(\theta)$ equals zero and the Fourier transform $U(\exp(j\theta))$ equals zero in the regions where $S(\theta)$ is non-zero. Because $U(z)$ is a polynomial of degree $N - 1$ with real coefficients, it can have a maximum of $N - 1$ zeros on the unit circle of the z-plane, occurring in conjugated pairs. This implies that for

110

$\mathbf{u}^T \mathbf{R} \mathbf{u}$ to be zero, $S(\theta)$ must differ from zero for at most $N - 1$ values of θ, and therefore it must be a sum of delta functions, at positions

$$\theta = \pm\theta_1, \pm\theta_2, \ldots, \pm\theta_r,$$

with $2r \le N - 1$. The possibilities of zeros at $\pm\pi$ are not taken into account. The autocorrelation function is in this case

$$R(k) = \sum_{l=1}^{r} \frac{A_l^2}{2} \cos(k\theta_l), \quad k = -\infty, \ldots, +\infty. \tag{A.3}$$

This is the autocorrelation function of a signal consisting of r sinusoidal components with random amplitude and phase, and A_l is the amplitude of the lth sinusoidal signal component. This is the only case in which a signal's autocorrelation matrix is really singular. However, there are other cases in which the autocorrelation matrix is almost singular. This is particularly so if $R(k)$ is the autocorrelation matrix of a band-limited signal $\underline{s}_j$, $j = -\infty, \ldots, +\infty$. If, for signal is low-pass, then (A.2) can be made small by choosing $\mathbf{u}$ such that $|U(\exp(j\theta))|$ is very small in the region where $S(\theta)$ is non-zero. Functions ideally suited for this purpose are the discrete prolate spheroidal wave functions, [50]. Slightly less well-suited to this purpose, but illustrative as an example, is a $U(\exp(j\theta))$ that is the transfer function of a high-pass filter that has a maximum attenuation in the pass-band of $\underline{s}_j$. There is a vast amount of signal processing literature on the design of filters that can do that very well, e.g. [48,49]. A stop-band attenuation of a filter of length 40, for instance, can easily be in the range of 60 dB [49], so that for this example a $\mathbf{u}$ can be found such that

$$\frac{\mathbf{u}^T \mathbf{R} \mathbf{u}}{\mathbf{u}^T \mathbf{u}} < 10^{-6} \, R(0).$$

An indication of the singularity of a matrix is its *condition number* κ, which, for positive definite symmetric matrices, can be defined by [22]

$$\kappa = \frac{\lambda_{\max}}{\lambda_{\min}}. \tag{A.4}$$

Here $\lambda_{\max}$ and $\lambda_{\min}$ are respectively the maximum and minimum eigenvalues of the matrix. If κ is high, the matrix is almost singular. Since, for $\mathbf{R}$,

$$\lambda_{\max} = \max_{\|\mathbf{u}\|^2=1} \mathbf{u}^T \mathbf{R} \mathbf{u} \ge R(0),$$

and

$$\lambda_{\min} = \min_{\|\mathbf{u}\|^2=1} \mathbf{u}^T \mathbf{R} \mathbf{u},$$

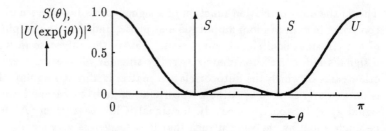

Figure A.1: Example of a spectrum $S(\exp(j\theta))$ and a Fourier transform $U(\exp(j\theta))$ belonging to a signal with a singular autocorrelation matrix.

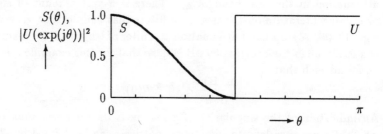

Figure A.2: Example of a spectrum $S(\exp(j\theta))$ and a transfer function $U(\exp(j\theta))$ belonging to a signal with an almost singular autocorrelation matrix.

for κ in the example above, where $U(\exp(j\theta))$ is the transfer function of a filter with a stop-band attenuation of 60 dB, one has $\kappa > 10^6$, which is generally considered to be very high. In these cases it is sometimes more convenient to treat the finite-sized autocorrelation matrix as singular. Figures A.1 and A.2 show examples of signal spectra and transfer functions $U(\exp(j\theta))$ for a singular and an almost singular autocorrelation matrix, respectively.

Appendix B

Properties of estimators

First it is shown that an $m \times N$ matrix $\mathbf{H}$ satisfying (2.5) and (2.7) always exists. From (2.7) it follows that the m rows of $\mathbf{H}$ must belong to the *null space* $\mathcal{N}(\mathbf{R}')$ of $\mathbf{R}'$ [22]. If $\mathbf{R}'$ has full rank, then $\mathcal{N}(\mathbf{R}')$ has dimension m, otherwise it has a dimension greater than m. For the null spaces of $\mathbf{R}'$ and $\mathbf{R}$ one has

$$\mathcal{N}(\mathbf{R}) \subset \mathcal{N}(\mathbf{R}').$$

Assume that $\mathcal{N}(\mathbf{R}')$ has dimension $n \geq m$. Let the rows of the $n \times N$ matrix $\mathbf{G}$ be a basis of $\mathcal{N}(\mathbf{R}')$.[1] If $\mathbf{H}$ exists, its rows are linear combinations of the rows of $\mathbf{G}$. This means that equivalent to the existence of $\mathbf{H}$ is the existence of an $m \times n$ matrix $\mathbf{A}$, such that

$$\mathbf{H} = \mathbf{A}\,\mathbf{G}.$$

Define the $n \times m$ matrix $\tilde{\mathbf{G}}$ by

$$\tilde{g}_{i,j} = g_{i,t(j)}, \quad i = 1, \ldots, n, \ j = 1, \ldots, m.$$

The matrix $\mathbf{A}$ must satisfy

$$\mathbf{A}\,\tilde{\mathbf{G}} = -\mathbf{I},$$

where $\mathbf{I}$ is the $m \times m$ identity matrix. This system of equations has solutions for $\mathbf{A}$ if $\mathrm{rank}(\tilde{\mathbf{G}}) = m$. It is shown that always $\mathrm{rank}(\tilde{\mathbf{G}}) = m$ and therefore, $\mathbf{A}$, and equivalently $\mathbf{H}$, exist. Suppose that

$$\mathrm{rank}(\tilde{\mathbf{G}}) < m.$$

---

[1]Note that the sizes of the matrices $\mathbf{G}$ and $\tilde{\mathbf{G}}$ in this appendix are different from those of the matrices $\mathbf{G}$ and $\tilde{\mathbf{G}}$ introduced in Section 2.2.

Then there is a vector $\tilde{\mathbf{w}} \in \mathbb{R}^m$, $\tilde{\mathbf{w}} \neq \mathbf{0}$, such that

$$\tilde{\mathbf{G}}\tilde{\mathbf{w}} = \mathbf{0}.$$

Define the N vector $\mathbf{w}$ by

$$\begin{aligned} w_{t(i)} &= \tilde{w}_i, \quad i = 1, \ldots, m, \\ w_j &= 0, \quad j \in W \setminus V. \end{aligned}$$

Obviously, $\mathbf{G}\mathbf{w} = \mathbf{0}$, and $\mathbf{w} \in \mathcal{N}(\mathbf{R}')^{\perp}$. Since [22, page 5]

$$\mathcal{N}(\mathbf{R}')^{\perp} = \mathcal{R}(\mathbf{R}'^T),$$

it follows, by using the fact that $\mathbf{R}$ is symmetric, that

$$\mathbf{w} \in \mathcal{R}(\mathbf{R}'^T) \subset \mathcal{R}(\mathbf{R}^T) = \mathcal{R}(\mathbf{R}).$$

Therefore, for some $\mathbf{y} \in \mathbb{R}^N$

$$\mathbf{R}\mathbf{y} = \mathbf{w} \neq \mathbf{0}.$$

Since $w_j = 0$, $j \in W \setminus V$, it follows that $\mathbf{R}'\mathbf{y} = \mathbf{0}$, or equivalently, $\mathbf{y} \in \mathcal{N}(\mathbf{R}')$, but now there is a contradiction, because $\mathbf{y} \in \mathcal{N}(\mathbf{R}')$ and $\mathbf{w} \in \mathcal{N}(\mathbf{R}')^{\perp}$, but

$$\mathbf{y}^T\mathbf{w} = \mathbf{y}^T\mathbf{R}\mathbf{y} > 0,$$

since $\mathbf{R}\mathbf{y} \neq \mathbf{0}$ and $\mathbf{R}$ is non-negative definite. This means that the assumption $\mathrm{rank}(\tilde{\mathbf{G}}) < m$ is wrong and that a set of weighting coefficients can always be found.

Some other interesting properties of the solutions of the system (2.7) can be derived. Let

$$\mathbf{G} = \begin{bmatrix} \mathbf{g}_1^T \\ \vdots \\ \mathbf{g}_n^T \end{bmatrix},$$

and

$$\tilde{\mathbf{G}} = \begin{bmatrix} \tilde{\mathbf{g}}_1^T \\ \vdots \\ \tilde{\mathbf{g}}_n^T \end{bmatrix}.$$

The rows of $\mathbf{G}$ belong either to $\mathcal{N}(\mathbf{R})$ or the subset T of $\mathbb{R}^N$, defined in (2.14) Assume, without loss of generality, that

$$\mathbf{g}_1, \ldots, \mathbf{g}_{m'} \in T,$$

and that

$$\mathbf{g}_{m'+1}, \ldots, \mathbf{g}_n \in \mathcal{N}(\mathbf{R}).$$

First it is shown that $m' \leq m$. Since $\mathcal{N}(\mathbf{R}) \subset \mathcal{N}(\mathbf{R}')$, the vectors $\mathbf{g}_{m'+1}, \ldots, \mathbf{g}_n$ are a basis for $\mathcal{N}(\mathbf{R})$. Now, clearly

$$\mathrm{rank}(\mathbf{R}) \leq \mathrm{rank}(\mathbf{R}') + m,$$

so that

$$N - (n - m') \leq (N - n) + m,$$

or

$$m' \leq m.$$

Furthermore, the first m' rows of $\tilde{\mathbf{G}}$ have full rank m'. The proof is as follows. Assume that the first m' rows of $\tilde{\mathbf{G}}$ have a rank less than m'. Then there is at least one linear combination $\mathbf{g}'$ of $\mathbf{g}_1, \ldots, \mathbf{g}_{m'}$ with

$$g'_{t(i)} = 0, \quad i = 1, \ldots, m.$$

Since $\mathbf{g}' \in T$, this implies that

$$\mathbf{g}'^T \mathbf{R} \mathbf{g}' = 0.$$

This is a contradiction, because $\mathbf{R}$ is non-negative definite and $\mathbf{R}\mathbf{g}' \neq \mathbf{0}$. Therefore the assumption that the first m' rows of $\tilde{\mathbf{G}}$ are not of full rank is false.

Also, if $m' > 0$, then there are precisely $m - m'$ independent vectors in $\tilde{\mathbf{g}}_{m'+1}, \ldots, \tilde{\mathbf{g}}_n$. To prove this, assume that there are $m'' > m - m'$ independent vectors in $\tilde{\mathbf{g}}_{m'+1}, \ldots, \tilde{\mathbf{g}}_n$. Without loss of generality it can be assumed that those vectors are $\tilde{\mathbf{g}}_{m'+1}, \ldots, \tilde{\mathbf{g}}_{m'+m''}$. Furthermore, note that if $\mathbf{g}_i \in T$ and $\mathbf{g}_j \in \mathcal{N}(\mathbf{R})$ then also $\mathbf{g}_i + \mathbf{g}_j \in T$. Now there is at least one linear combination $\mathbf{g}' \in T$ of vectors $\mathbf{g}_1, \ldots, \mathbf{g}_{m'}$ and $\mathbf{g}_{m'+1}, \ldots, \mathbf{g}_{m'+m''}$ such that

$$g'_{t(j)} = 0, \quad j = 1, \ldots, m.$$

But then,

$$\mathbf{g}'^T \mathbf{R} \mathbf{g}' = 0.$$

This is not true because $\mathbf{g}' \in T$, and therefore the assumption that there are more than $m - m'$ independent vectors in $\tilde{\mathbf{g}}_{m'+1}, \ldots, \tilde{\mathbf{g}}_n$ is false.

Appendix C

The restoration error

In this appendix[1] the upper bound (3.12)

$$\frac{\sigma_e^2}{b_0 R(0)} \leq \frac{1 - (\prod_{i=1}^{p} \rho_i)^2}{1 + (\prod_{i=1}^{p} \rho_i)^2}$$

for the relative restoration error variance is derived. This upper bound is valid for the case of one unknown sample and an autoregressive process of order p.

Note that since the order of the autoregressive process is p, one has that $a_p \neq 0$. For $b_0 = 1 + a_1^2 + \ldots + a_p^2$ (3.3) it follows straightforwardly that

$$b_0 \geq 1 + a_p^2. \tag{C.1}$$

A similar expression for $R(0)/\sigma_e^2$ is more difficult to obtain. It comes in fact as a by-product of the Levinson-Durbin [33,30,31] algorithm. On multiplying both sides of (3.1) by $\underline{s}_{k-m}$, $m = 0, \ldots, p$, and by taking the expected value, one obtains after some manipulations the *Yule-Walker* equations [31]:

$$\mathbf{R}^{(p)} \mathbf{a} = -\mathbf{r}^{(p)}. \tag{C.2}$$

Here $\mathbf{R}^{(p)}$ is the $p \times p$ autocorrelation matrix, defined by

$$r_{i,j}^{(p)} = R(i - j), \ i, j = 1, \ldots, p,$$

$\mathbf{r}^{(p)}$ is a vector of autocorrelation coefficients, defined by

$$\mathbf{r}^{(p)} = [R(1), \ldots, R(p)]^T,$$

and $\mathbf{a}$ is the vector of prediction coefficients, defined by

$$\mathbf{a} = [a_1, \ldots, a_p]^T.$$

---

[1]The results of this appendix have been published previously in [55].

117

The Levinson-Durbin algorithm is an iterative procedure to solve (C.2) for $\mathbf{a}$ in p iteration steps [31]. In the ith iteration step the system

$$\mathbf{R}^{(i)}\mathbf{a}^{(i)} = -\mathbf{r}^{(i)}$$

is solved for $\mathbf{a}^{(i)}$ by using the already obtained $\mathbf{a}^{(i-1)}$. Here $\mathbf{R}^{(i)}$ and $\mathbf{r}^{(i)}$ are defined as above, with p replaced by i. The Levinson-Durbin algorithm also computes σ_e^2 in p iteration steps. Starting from $(\sigma_e^2)^{(0)} = R(0)$, in the ith iteration step $(\sigma_e^2)^{(i)}$ is computed by

$$(\sigma_e^2)^{(i)} = (1 - (a_i^{(i)})^2)(\sigma_e^2)^{(i-1)}. \tag{C.3}$$

The coefficient $a_i^{(i)}$ is often called the ith *reflection coefficient*. For a polynomial $A(z)$ (3.5) of order p, with all its zeros inside the unit circle of the complex plane, one has that

$$|a_i^{(i)}| < 1, \; i = 1, \ldots, p.$$

The result of the Levinson-Durbin algorithm are sequences

$$1, \mathbf{a}^{(1)}, \ldots, \mathbf{a}^{(p)},$$

and

$$R(0), (\sigma_e^2)^{(1)}, \ldots, (\sigma_e^2)^{(p)},$$

with $\mathbf{a}^{(p)} = \mathbf{a}$ and $(\sigma_e^2)^{(p)} = \sigma_e^2$. By repeatedly using (C.3) one obtains

$$\frac{R(0)}{\sigma_e^2} = \frac{1}{(1 - (a_1^{(1)})^2)(1 - (a_2^{(2)})^2)\ldots(1 - (a_p^{(p)})^2)}.$$

Because $a_p^{(p)} = a_p \neq 0$ and $|a_i^{(i)}| < 1$, this gives

$$\frac{R(0)}{\sigma_e^2} \geq \frac{1}{1 - a_p^2}. \tag{C.4}$$

The polynomial $A(z)$ in (3.5) can also be written as

$$A(z) = (1 - \alpha_1 z^{-1})(1 - \alpha_2 z^{-1})\ldots(1 - \alpha_p z^{-1}),$$

where the α_i, $i = 1, \ldots, p$, are the complex zeros. For a_p it follows that

$$a_p = \alpha_1 \alpha_2 \ldots \alpha_p.$$

Because $A(z)$ has real coefficients, the zeros occur in conjugated pairs and, if $\alpha_i = \rho_i \exp(j\phi_i)$, then

$$a_p = \rho_1 \rho_2 \ldots \rho_p. \tag{C.5}$$

Combining (C.1), (C.4) and (C.5) gives

$$\frac{\sigma_e^2}{b_0 R(0)} \leq \frac{1 - (\prod_{i=1}^p \rho_i)^2}{1 + (\prod_{i=1}^p \rho_i)^2},$$

which proves the claim.

The polynomial

$$1 - a_p z^{-p}$$

has zeros

$$\alpha_k = |a_p|^{\frac{1}{p}} \exp(j\frac{2\pi}{p}k), \ k = 1, \ldots, p,$$

if $a_p > 0$, or

$$\alpha_k = |a_p|^{\frac{1}{p}} \exp(j(\frac{2\pi}{p}k + \frac{\pi}{p})), \ k = 1, \ldots, p,$$

if $a_p < 0$. It is obvious that for this polynomial (C.1) holds with the equality sign. It follows from the Levinson-Durbin algorithm that

$$a_i^{(i)} = 0, \ i = 1, \ldots, p - 1,$$

if and only if

$$a_i = 0, \ i = 1, \ldots, p - 1,$$

therefore (C.4) also holds with the equality sign. Consequently, (3.12) holds with the equality sign.

Appendix D

Analysis of $Q(\mathbf{a}, \mathbf{x})$

D.1 Statistical parameter estimation

It is shown that, under the assumption that the signal $\underline{s}_k$ has a Gaussian probability density function, the log likelihood function (3.15) is maximized as a function of $\mathbf{a}$ by minimizing $Q(\mathbf{a}, \mathbf{x})$ for known $\mathbf{x}$.

It is assumed that the $\underline{e}_k$, $k = -\infty, \ldots, +\infty$, (3.1) are independent and have probability density functions

$$p_{\underline{e}_k}(e) = \frac{1}{\sigma_e \sqrt{2\pi}} \exp\left(-\frac{e^2}{2\sigma_e^2}\right), \quad k = -\infty, \ldots, +\infty. \qquad \text{(D.1)}$$

The log likelihood function that is usually used to obtain maximum likelihood estimates for σ_e^2 and $\mathbf{a}$ from a sequence $\mathbf{s} = [s_1, \ldots, s_N]^T$ is $\log(p_{\underline{\mathbf{s}}}(\mathbf{s}|\sigma_e^2, \mathbf{a}))$, the logarithm of the joint probability density function of $\mathbf{s}$. The log likelihood function in (3.15) differs slightly from this one. However, it can be shown that for large N, compared to p, one may approximate the more commonly used log likelihood function by the one given in (3.15) [56].

To express $L(\sigma_e^2, \mathbf{a})$ in terms of $Q(\mathbf{a}, \mathbf{x})$ one observes that

$$
\begin{aligned}
& p_{\underline{\mathbf{s}}|\underline{\mathbf{u}}}(\mathbf{s}|\mathbf{u}, \sigma_e^2, \mathbf{a}) \\
&= p_{\underline{s}_{p+1}, \ldots, \underline{s}_N | \underline{s}_1, \ldots, \underline{s}_p}(s_{p+1}, \ldots, s_N | s_1, \ldots, s_p, \sigma_e^2, \mathbf{a}) \\
&= p_{\underline{s}_{p+2}, \ldots, \underline{s}_N | \underline{s}_1, \ldots, \underline{s}_{p+1}}(s_{p+2}, \ldots, s_N | s_1, \ldots, s_{p+1}, \sigma_e^2, \mathbf{a}) \\
& \quad \times p_{\underline{s}_{p+1} | \underline{s}_1, \ldots, \underline{s}_p}(s_{p+1} | s_1, \ldots, s_p, \sigma_e^2, \mathbf{a}).
\end{aligned}
$$

Furthermore,

$$p_{\underline{s}_{p+1} | \underline{s}_1, \ldots, \underline{s}_p}(s_{p+1} | s_1, \ldots, s_p, \sigma_e^2, \mathbf{a})$$

120

$$= p_{\underline{\varepsilon}_{p+1}}\left(\sum_{l=0}^{p} a_l s_{p+1-l}\right)$$

$$= \frac{1}{\sigma_e\sqrt{2\pi}}\exp\left(-\frac{1}{2\sigma_e^2}\left(\sum_{l=0}^{p} a_l s_{p+1-l}\right)^2\right).$$

By repeatedly applying the above reasoning, one finds that

$$p_{\underline{s}|\underline{u}}(s|u,\sigma_e^2,\mathbf{a}) = \left(\frac{1}{\sigma_e\sqrt{2\pi}}\right)^{N-p}\exp\left(-\frac{1}{2\sigma_e^2}Q(\mathbf{a},\mathbf{x})\right). \tag{D.2}$$

Therefore,

$$L(\sigma_e^2,\mathbf{a}) = -(N-p)\log(\sigma_e\sqrt{2\pi}) - \frac{1}{2\sigma_e^2}Q(\mathbf{a},\mathbf{x}). \tag{D.3}$$

Maximizing $L(\sigma_e^2,\mathbf{a})$ as a function of $\mathbf{a}$ is the same as minimizing $Q(\mathbf{a},\mathbf{x})$ as a function of $\mathbf{a}$. This proves the claim. Furthermore, σ_e^2 can be estimated by maximizing $L(\sigma_e^2,\mathbf{a})$ as a function of σ_e^2. This gives the estimate of (3.14) if $m=0$ is taken.

D.2 Statistical sample estimation

It is shown that, under the hypothesis that the excitation noise samples have probability density functions (D.1), finding the minimum variance estimate for $\mathbf{x}$, with given $\mathbf{a}$ and s_k, $k = 1,\ldots,N$, $k \neq t(1),\ldots,t(m)$, is the same as minimizing $Q(\mathbf{a},\mathbf{x})$ as a function of $\mathbf{x}$. To this end one can use the well-known fact from statistical estimation theory that the minimum variance estimator $\hat{\underline{\mathbf{x}}}_{\mathrm{mv}}$ of $\underline{\mathbf{x}}$, given $\underline{\mathbf{v}}$, follows from [21]

$$\hat{\underline{\mathbf{x}}}_{\mathrm{mv}} = \mathcal{E}\left\{\underline{\mathbf{x}}|\underline{\mathbf{v}}\right\}.$$

It is straightforward to show that

$$p_{\underline{\mathbf{x}}|\underline{\mathbf{v}}}(\mathbf{x}|\mathbf{v}) = \frac{p_{\underline{s}|\underline{u}}(s|u)}{p_{\underline{\mathbf{v}}|\underline{u}}(\mathbf{v}|\mathbf{u})}. \tag{D.4}$$

By using (D.2), one can express the right-hand side of (D.4) in terms of $Q(\mathbf{a},\mathbf{x})$. More specifically, one has for $\hat{\underline{\mathbf{x}}}_{\mathrm{mv}}$

$$\hat{\underline{\mathbf{x}}}_{\mathrm{mv}} = D\int_{\mathbf{x}\in\mathbb{R}^m}\mathbf{x}\exp\left(-\frac{1}{2\sigma_e^2}Q(\mathbf{a},\mathbf{x})\right)d\mathbf{x}, \tag{D.5}$$

where D is a constant such that

$$D \int_{x \in \mathbb{R}^m} \exp \left(-\frac{1}{2\sigma_e^2} Q(a, x) \right) dx = 1.$$

It follows from a standard fact about Gaussian integrals that the function $Q(a, x)$, a quadratic form in x, is minimized by $\hat{\underline{x}}_{mv}$ in (D.5). This proves the claim.

Since $\hat{\underline{x}}_{mv}$ also maximizes $p_{\underline{x}|\underline{v}}(x|v)$ as a function of x, it is also a maximum a posteriori estimate for x. This follows from (D.4) and (D.2).

Appendix E

Minimization of $Q(\mathbf{a}, \mathbf{x})$

E.1 Convergence properties

Iterating the restoration method of Chapter 3 comes down to constructing two sequences $\hat{\mathbf{a}}^{(k)}$, $\hat{\mathbf{x}}^{(k)}$ of vectors of prediction coefficients and sample estimates, respectively. In the kth step $\hat{\mathbf{a}}^{(k)}$ and $\hat{\mathbf{x}}^{(k)}$ are obtained by minimizing $Q(\mathbf{a}, \hat{\mathbf{x}}^{(k-1)})$ with respect to $\mathbf{a}$ and $Q(\hat{\mathbf{a}}^{(k)}, \mathbf{x})$ with respect to $\mathbf{x}$, respectively. That is

$$Q(\hat{\mathbf{a}}^{(k)}, \hat{\mathbf{x}}^{(k-1)}) = \min_{\mathbf{a} \in \mathbb{R}^p} Q(\mathbf{a}, \hat{\mathbf{x}}^{(k-1)}),$$

$$Q(\hat{\mathbf{a}}^{(k)}, \hat{\mathbf{x}}^{(k)}) = \min_{\mathbf{x} \in \mathbb{R}^m} Q(\hat{\mathbf{a}}^{(k)}, \mathbf{x}).$$

It is found that iterating the method improves the results if the number of available samples is relatively small. Although the method converges rapidly in practice, it does not seem easy to prove satisfactory convergence results. It can be shown that, when the sequence $(\hat{\mathbf{a}}^{(k)}, \hat{\mathbf{x}}^{(k)})$ converges, the limit point $(\hat{\mathbf{a}}^{(\infty)}, \hat{\mathbf{x}}^{(\infty)})$ is a stationary point. However, $Q(\mathbf{a}, \mathbf{x})$ may have several of such points. For the asymptotic speed of convergence, the Hessian $\mathbf{H}^{(\infty)}$ at $(\hat{\mathbf{a}}^{(\infty)}, \hat{\mathbf{x}}^{(\infty)})$ is relevant. Letting

$$\mathbf{H}^{(k)} = \begin{bmatrix} \mathbf{A}_{1,1}^{(k)} & \mathbf{A}_{1,2}^{(k)} \\ (\mathbf{A}_{1,2}^{(k)})^T & \mathbf{A}_{2,2}^{(k)} \end{bmatrix}, \tag{E.1}$$

where

$$\mathbf{A}_{1,1}^{(k)} = \frac{\partial^2 Q(\hat{\mathbf{a}}^{(k)}, \hat{\mathbf{x}}^{(k)})}{\partial \mathbf{a}^2},$$

$$\mathbf{A}_{1,2}^{(k)} = \frac{\partial^2 Q(\hat{\mathbf{a}}^{(k)}, \hat{\mathbf{x}}^{(k)})}{\partial \mathbf{a} \partial \mathbf{x}},$$

$$\mathbf{A}_{2,2}^{(k)} = \frac{\partial^2 Q(\hat{\mathbf{a}}^{(k)}, \hat{\mathbf{x}}^{(k)})}{\partial \mathbf{x}^2},$$

it follows that

$$
\begin{aligned}
\mathbf{A}_{1,1}^{(k)} &= 2\mathbf{C}^{(k)}, \\
\left(\mathbf{A}_{1,2}^{(k)}\right)_{i,j} &= 2(e_{i+t(j)}^{(k)} + f_{-i+t(j)}^{(k)}), \quad i = 1,\ldots,p, \; j = 1,\ldots,m, \\
\mathbf{A}_{2,2}^{(k)} &= 2\tilde{\mathbf{G}}^{(k)}.
\end{aligned}
$$

Here $\mathbf{C}^{(k)}$ and $\tilde{\mathbf{G}}^{(k)}$ are the matrices $\mathbf{C}$ and $\tilde{\mathbf{G}}$ from (3.17) and (3.8), respectively, computed with $(\mathbf{a},\mathbf{x}) = (\hat{\mathbf{a}}^{(k)}, \hat{\mathbf{x}}^{(k)})$, and, with $s_{t(i)} = \hat{x}_i^{(k)}$, $i = 1,\ldots,m$,

$$
\begin{aligned}
e_j^{(k)} &= \sum_{l=0}^{p} a_l^{(k)} s_{j-l}, \\
f_j^{(k)} &= \sum_{l=0}^{p} a_l^{(k)} s_{j+l}.
\end{aligned}
$$

In the neighbourhood of $(\hat{\mathbf{a}}^{(\infty)}, \hat{\mathbf{x}}^{(\infty)})$ $Q(\mathbf{a},\mathbf{x})$ is given by

$$
Q(\mathbf{a},\mathbf{x}) \cong
$$

$$
Q(\hat{\mathbf{a}}^{(\infty)}, \hat{\mathbf{x}}^{(\infty)}) + \frac{1}{2}(\mathbf{a} - \hat{\mathbf{a}}^{(\infty)})^T \mathbf{A}_{1,1}(\mathbf{a} - \hat{\mathbf{a}}^{(\infty)})
$$

$$
+ (\mathbf{a} - \hat{\mathbf{a}}^{(\infty)})^T \mathbf{A}_{1,2}(\mathbf{x} - \hat{\mathbf{x}}^{(\infty)}) + \frac{1}{2}(\mathbf{x} - \hat{\mathbf{x}}^{(\infty)})^T \mathbf{A}_{2,2}(\mathbf{x} - \hat{\mathbf{x}}^{(\infty)}).
$$

It follows that

$$
\hat{\mathbf{x}}^{(k)} \cong \hat{\mathbf{x}}^{(\infty)} + \mathbf{A}_{2,2}^{-1} \mathbf{A}_{1,2}^T \mathbf{A}_{1,1}^{-1} \mathbf{A}_{1,2}(\hat{\mathbf{x}}^{(k-1)} - \hat{\mathbf{x}}^{(\infty)}).
$$

The speed of convergence is determined by the eigenvalues of

$$
\mathbf{D} = \mathbf{A}_{2,2}^{-1} \mathbf{A}_{1,2}^T \mathbf{A}_{1,1}^{-1} \mathbf{A}_{1,2}.
$$

A condition guaranteeing linear convergence is that eigenvalues of $\mathbf{D}^T \mathbf{D}$ are all less than 1. It does not seem easy to check this condition.

E.2 The EM algorithm

Assume that the excitation noise samples have probability density functions (D.1). Consider the log likelihood function (D.3). The EM algorithm aims to find estimates for parameters and unknown samples from incomplete data by maximizing the (log) likelihood function [35,36,37]. For the present situation it can be described as follows. Starting with initial estimates $(\hat{\sigma}_e^2)^{(0)}$, $\hat{\mathbf{a}}^{(0)}$, one constructs sequences $(\hat{\sigma}_e^2)^{(k)}$, $\hat{\mathbf{a}}^{(k)}$, $k = 1, 2, \ldots$, by choosing in the kth step $(\hat{\sigma}_e^2)^{(k)}$, $\hat{\mathbf{a}}^{(k)}$, in such a way that

$$\mathcal{E}\left\{ L(\sigma_e^2, \mathbf{a}) | \underline{\mathbf{v}}, (\hat{\sigma}_e^2)^{(k-1)}, \hat{\mathbf{a}}^{(k-1)} \right\}$$

is maximal at

$$(\sigma_e^2, \mathbf{a}) = ((\hat{\sigma}_e^2)^{(k)}, \hat{\mathbf{a}}^{(k)}).$$

One would like to maximize $L(\sigma_e^2, \mathbf{a})$, but this is impossible, since one does not know s completely.

To show the connection with the iterative restoration method, it is necessary to evaluate

$$\mathcal{E}\left\{ L(\tilde{\sigma}_e^2, \tilde{\mathbf{a}}) | \underline{\mathbf{v}}, \sigma_e^2, \mathbf{a} \right\}.$$

The conditional expectation in this expression refers to the conditional probability density function

$$p_{\underline{\mathbf{x}}|\underline{\mathbf{v}}}(\mathbf{x}|\mathbf{v}) = \frac{|\tilde{\mathbf{G}}|^{\frac{1}{2}}}{(2\pi)^{\frac{m}{2}} \sigma_e^m} \exp\left(-\frac{(\mathbf{x} - \hat{\mathbf{x}})^T \tilde{\mathbf{G}}(\mathbf{x} - \hat{\mathbf{x}})}{2\sigma_e^2} \right).$$

Here $\hat{\mathbf{x}}$ is the linear minimum variance estimate of Chapter (3). It follows that

$$\mathcal{E}\left\{ L(\tilde{\sigma}_e^2, \tilde{\mathbf{a}}) | \underline{\mathbf{v}}, \sigma_e^2, \mathbf{a} \right\} =$$
$$-(N - p)\log(\tilde{\sigma}_e \sqrt{2\pi}) - \frac{\mathcal{E}\left\{ Q(\tilde{\mathbf{a}}, \underline{\mathbf{x}}) \mid \underline{\mathbf{v}}, \sigma_e^2, \mathbf{a} \right\}}{2\tilde{\sigma}_e^2}.$$

It is a tedious but straightforward calculation to show that

$$\mathcal{E}\left\{ Q(\tilde{\mathbf{a}}, \underline{\mathbf{x}}) \mid \underline{\mathbf{v}}, \sigma_e^2, \mathbf{a} \right\} = \tag{E.2}$$
$$\sigma_e^2 \mathrm{trace}\left(\left(\tilde{\mathbf{G}}(\mathbf{a})\right)^{-1} \tilde{\mathbf{G}}(\tilde{\mathbf{a}}) \right) + Q(\tilde{\mathbf{a}}, \hat{\mathbf{x}}).$$

Here $\tilde{\mathbf{G}}(\mathbf{a})$ and $\tilde{\mathbf{G}}(\tilde{\mathbf{a}})$ denote that $\tilde{\mathbf{G}}$ from (3.8) is computed with prediction coefficients $\mathbf{a}$ and $\tilde{\mathbf{a}}$, respectively. Combining the results, one finds

$$\mathcal{E}\left\{ L(\tilde{\sigma}_e^2, \tilde{\mathbf{a}}) | \underline{\mathbf{v}}, \sigma_e^2, \mathbf{a} \right\} =$$
$$-(N - p)\log(\tilde{\sigma}_e \sqrt{2\pi}) - \frac{\sigma_e^2}{2\tilde{\sigma}_e^2} \mathrm{trace}\left(\left(\tilde{\mathbf{G}}(\mathbf{a})\right)^{-1} \tilde{\mathbf{G}}(\tilde{\mathbf{a}}) \right)$$
$$-\frac{Q(\tilde{\mathbf{a}}, \hat{\mathbf{x}})}{2\tilde{\sigma}_e^2}.$$

Maximizing $\mathcal{E}\left\{L(\tilde{\sigma}_e^2, \tilde{\mathbf{a}})|\underline{\mathbf{y}}, \sigma_e^2, \mathbf{a}\right\}$ is the same as minimizing (E.2). Hence the difference between the EM algorithm and the restoration method is reflected by the first term on the right-hand side of (E.2). It should be noted that the minimization of the right-hand side of (E.2) is more complicated because of the presence of this first term.

E.3   Newton-Raphson's method

Minimization of $Q(\mathbf{a}, \mathbf{x})$ with Newton-Raphson's method [39] comes down to the construction of a sequence of vectors

$$\begin{bmatrix} \hat{\mathbf{a}}^{(k)} \\ \hat{\mathbf{x}}^{(k)} \end{bmatrix}, \ k = 0, 1, \ldots,$$

of which the kth vector is computed from the $(k-1)$th by

$$\begin{bmatrix} \hat{\mathbf{a}}^{(k)} \\ \hat{\mathbf{x}}^{(k)} \end{bmatrix} = \begin{bmatrix} \hat{\mathbf{a}}^{(k-1)} \\ \hat{\mathbf{x}}^{(k-1)} \end{bmatrix} -$$

$$\begin{bmatrix} \frac{\partial^2 Q(\hat{\mathbf{a}}^{(k-1)}, \hat{\mathbf{x}}^{(k-1)})}{\partial \mathbf{a}^2} & \frac{\partial^2 Q(\hat{\mathbf{a}}^{(k-1)}, \hat{\mathbf{x}}^{(k-1)})}{\partial \mathbf{a} \partial \mathbf{x}} \\ \frac{\partial^2 Q(\hat{\mathbf{a}}^{(k-1)}, \hat{\mathbf{x}}^{(k-1)})}{\partial \mathbf{x} \partial \mathbf{a}} & \frac{\partial^2 Q(\hat{\mathbf{a}}^{(k-1)}, \hat{\mathbf{x}}^{(k-1)})}{\partial \mathbf{x}^2} \end{bmatrix}^{-1} \begin{bmatrix} \frac{\partial Q(\hat{\mathbf{a}}^{(k-1)}, \hat{\mathbf{x}}^{(k-1)})}{\partial \mathbf{a}} \\ \frac{\partial Q(\hat{\mathbf{a}}^{(k-1)}, \hat{\mathbf{x}}^{(k-1)})}{\partial \mathbf{a}} \end{bmatrix},$$

or, by using (E.1),

$$\begin{bmatrix} \hat{\mathbf{a}}^{(k)} \\ \hat{\mathbf{x}}^{(k)} \end{bmatrix} = \begin{bmatrix} \hat{\mathbf{a}}^{(k-1)} \\ \hat{\mathbf{x}}^{(k-1)} \end{bmatrix} \tag{E.3}$$

$$- \left(\mathbf{H}^{(k-1)}\right)^{-1} \begin{bmatrix} 2\mathbf{C}^{(k-1)}\hat{\mathbf{a}}(k-1) + 2\mathbf{c}^{(k-1)} \\ 2\tilde{\mathbf{G}}^{(k-1)}\hat{\mathbf{x}}(k-1) + 2\mathbf{z}^{(k-1)} \end{bmatrix}.$$

Here $\mathbf{c}^{(k-1)}$ and $\mathbf{z}^{(k-1)}$ are computed with $\hat{\mathbf{x}}^{(k-1)}$ and $\hat{\mathbf{a}}^{(k-1)}$. The relation with the restoration method of Chapter 3 becomes evident when $\mathbf{H}^{(k-1)}$ in (E.3) is modified by setting

$$\mathbf{A}_{1,2}^{(k-1)} = \mathbf{0}.$$

As a result, one obtains

$$\begin{bmatrix} \hat{\mathbf{a}}^{(k)} \\ \hat{\mathbf{x}}^{(k)} \end{bmatrix} = - \begin{bmatrix} \mathbf{C}^{(k-1)} & \mathbf{0} \\ \mathbf{0} & \tilde{\mathbf{G}}^{(k-1)} \end{bmatrix}^{-1} \begin{bmatrix} \mathbf{c}^{(k-1)} \\ \mathbf{z}^{(k-1)} \end{bmatrix}. \tag{E.4}$$

It is easily verified that, starting from initial estimates $\hat{\mathbf{a}}^{(0)} = \mathbf{0}$ and $\hat{\mathbf{x}}^{(0)} = \mathbf{0}$, a sequence

$$\begin{bmatrix} \mathbf{0} \\ \mathbf{0} \end{bmatrix}, \begin{bmatrix} \hat{\mathbf{a}}^{(1)} \\ \mathbf{0} \end{bmatrix}, \begin{bmatrix} \hat{\mathbf{a}}^{(1)} \\ \hat{\mathbf{x}}^{(1)} \end{bmatrix}, \begin{bmatrix} \hat{\mathbf{a}}^{(2)} \\ \hat{\mathbf{x}}^{(1)} \end{bmatrix}, \begin{bmatrix} \hat{\mathbf{a}}^{(2)} \\ \hat{\mathbf{x}}^{(2)} \end{bmatrix}, \ldots$$

is obtained, where the $\hat{\mathbf{a}}^{(k)}$, $\hat{\mathbf{x}}^{(k)}$ in this sequence are identical to those obtained with the algorithm described in Chapter 3.

Appendix F

Decomposition of $\tilde{\mathbf{G}}$

In this appendix the left-hand inequality of (3.28) is derived. First note that

$$\tilde{\mathbf{G}} = \mathbf{A}^T \mathbf{A}, \qquad (\text{F.1})$$

where $\mathbf{A} = [\mathbf{a}_1, \ldots, \mathbf{a}_m]$ is a $(t(m) - t(1) + p + 1) \times m$ matrix, defined by

$$a_{i,j} = (\mathbf{a}_j)_i = a_{t(j)-t(1)-i+p+1},$$

the a_i being the prediction coefficients, with $a_i = 0$ for $i < 0$ or $i > p$. Since $\mathbf{A}$ has full rank, $\mathbf{A} = \mathbf{QR}$, where $\mathbf{Q}$ is a $(t(m) - t(1) + p + 1) \times m$ matrix, consisting of m orthogonal columns $\mathbf{q}_i$, $i = 1, \ldots, m$, and $\mathbf{R}$ is an upper triangular $m \times m$ matrix. This decomposition is generally referred to as QR decomposition [22]. On substituting $\mathbf{A} = \mathbf{QR}$ into (F.1) one obtains

$$\tilde{\mathbf{G}} = \mathbf{R}^T \mathbf{Q}^T \mathbf{Q} \ \mathbf{R} = \tilde{\mathbf{L}}^T \mathbf{D} \tilde{\mathbf{L}},$$

where $\tilde{\mathbf{L}}$ and $\mathbf{D}$ are as in (3.27). Clearly, $d_{j,j} = \parallel \mathbf{q}_j \parallel^2$. The QR decomposition of $\mathbf{A}$ can be done iteratively. In the jth step, $\mathbf{q}_j$ is found by subtracting from $\mathbf{a}_j$ its projection on $\text{span}\{\mathbf{q}_1, \ldots, \mathbf{q}_{j-1}\}$:

$$\mathbf{q}_j = \mathbf{a}_j - \sum_{k=1}^{j-1} \frac{\mathbf{a}_j^T \mathbf{q}_k}{\parallel \mathbf{q}_j \parallel^2}.$$

Because

$$\text{span}\{\mathbf{q}_1, \ldots, \mathbf{q}_{j-1}\} = \text{span}\{\mathbf{a}_1, \ldots, \mathbf{a}_{j-1}\},$$

one has

$$
\begin{aligned}
\parallel \mathbf{q}_j \parallel^2 &= \min_{\mathbf{y} \in \text{span}\{\mathbf{q}_1, \ldots, \mathbf{q}_{j-1}\}} \parallel \mathbf{a}_j - \mathbf{y} \parallel^2 \\
&= \min_{\mathbf{y} \in \text{span}\{\mathbf{a}_1, \ldots, \mathbf{a}_{j-1}\}} \parallel \mathbf{a}_j - \mathbf{y} \parallel^2 \\
&= \min_{\mathbf{w} \in \mathbb{R}^{j-1}} \left(\mathbf{a}_j + \sum_{k=1}^{j-1} w_k \mathbf{a}_k \right)^2.
\end{aligned}
$$

Since $(a_j)_{t(j)-t(1)+p+1} = a_0 = 1$ and $(a_k)_{t(j)-t(1)+p+1} = 0$ for $k = 1, \ldots, j$, it follows easily that

$$\| \ q_j \ \|^2 \geq 1, \ j = 1, \ldots, m.$$

This proves the left-hand inequality of (3.28).

Appendix G

Two-dimensional predictors

To obtain the prediction coefficients $a_{k,l}$ the function

$$Q_{\mathbf{a}}(\mathbf{a}) = \sum_{\text{range}(i,j)} \left(\sum_{\text{range}(k,l)} a_{k,l} s_{i-k,j-l} \right)^2 \qquad (G.1)$$

is minimized as a function of the unknown prediction coefficients. For notational convenience, these are arranged into the vector $\mathbf{a}$. Since $a_{0,0} = 1$, (G.1) can be rewritten as

$$Q_{\mathbf{a}}(\mathbf{a}) = \sum_{\text{range}(i,j)} \left(s_{i,j} + \sum_{\text{range}(k,l),k,l \neq 0,0} a_{k,l} s_{i-k,j-l} \right)^2. \qquad (G.2)$$

Assume that $\mathbf{a}$ has elements

$$a_i = a_{k(i),l(i)}, \quad i = 1, \dots, r. \qquad (G.3)$$

Here r is the number of unknown prediction coefficients and the sequences $k(i), l(i)$, $i = 1, \dots, r$, determine how the prediction coefficients are mapped onto the elements of the vector $\mathbf{a}$. A formal introduction of the use of this kind of mappings to estimate two-dimensional prediction coefficients as one-dimensional vectors is given in [57]. Now (G.2) can be rewritten as

$$Q(\mathbf{a}) = \sum_{\text{range}(i,j)} \left(s_{i,j} + \sum_{n=1}^{r} a_n s_{i-k(n),j-l(n)} \right)^2. \qquad (G.4)$$

This function is now minimized as a function of $\mathbf{a}$ by computing the derivatives with respect to the elements of $\mathbf{a}$, setting these equal to zero and solving the resulting system of r linear equations with r unknowns

$$\mathbf{C}\hat{\mathbf{a}} = -\mathbf{c}. \qquad (G.5)$$

129

Here

$$c_{u,v} = \sum_{\text{range}(i,j)} s_{i-k(u),j-l(u)} s_{i-k(v),j-l(v)}, \qquad \text{(G.6)}$$

where the $r \times r$ matrix $\mathbf{C}$ has elements $c_{i,j}$, $i, j = 1, \ldots, r$, and the vector $\mathbf{c}$ is given by $\mathbf{c} = [c_{0,1}, \ldots, c_{0,r}]^T$. As is the case in [7], C and $\mathbf{c}$ both consist of autocovariance coefficients, but C is no longer structured in the same convenient way, and it is not clear whether efficient inversion algorithms for matrices of this type exist. This method is generally referred to as the *autocovariance method* [53]. If the pixel values outside the block are set equal to zero, and the range(i, j) is over all the integers, then C is a block-Toeplitz matrix, and the method is called the *autocorrelation method*. Fast system-solving algorithms do exist for this method [47].

References

[1] R.N.J. Veldhuis. *Adaptive Restoration of Lost Samples in Discrete-Time Signals and Digital Images*. PhD thesis, Katholieke Universiteit Nijmegen, June 1988.

[2] J.B.H. Peek. Communication aspects of the Compact Disc digital audio system. *IEEE Communications Magazine*, 23(2):7–15, 1985.

[3] A.J. Viterbi and J.K. Omura. *Principles of Digital Communications and Coding*. McGraw-Hill/Kogakusha, Tokyo, 1979.

[4] H. Hoeve, J. Timmermans, and L.B. Vries. Error correction and concealment in the Compact Disc. *Philips Technical Review*, (40):166–172, 1982.

[5] A.J.E.M. Janssen and L.B. Vries. Interpolation of band-limited discrete-time signals by minimizing out-of-band energy. In *Proceedings ICASSP-84*, San Diego, 1984.

[6] P. Delsarte, A.J.E.M. Janssen, and L.B. Vries. Discrete prolate spheroidal wave functions and interpolation. *SIAM J. Appl. Math.*, 45:641–650, 1985.

[7] A.J.E.M. Janssen, R.N.J. Veldhuis, and L.B. Vries. Adaptive interpolation of discrete-time signals that can be modeled as autoregressive processes. *IEEE Transactions on ASSP*, 34(2):317–330, 1986.

[8] R.N.J. Veldhuis, A.J.E.M. Janssen, and L.B. Vries. Adaptive restoration of unknown samples in certain time-discrete signals. In *Proceedings ICASSP-85*, pages 1013–1016, Tampa, 1985.

[9] J.L. Flanagan. *Speech Analysis Synthesis and Perception*. Springer-Verlag, Berlin, 1972.

131

[10] R.N.J. Veldhuis. A method for the restoration of burst errors in speech signals. In *Signal Processing 3: Theories and Applications*, pages 403–406, North-Holland, Amsterdam, 1986.

[11] R.N.J. Veldhuis and A.J.E.M. Janssen. A unified approach to the restoration of lost samples in discrete-time signals. In *Proceedings 21st Asilomar Conference*, Asilomar, 1986.

[12] N.S. Jayant and P. Noll. *Digital Coding of Waveforms*. Prentice Hall, Englewood Cliffs, New Jersey, 1984.

[13] R.W. Gerchberg. Super-resolution through error energy reduction. *Optica Acta*, 22:691–695, 1975.

[14] R.J. Marks II. Restoring lost samples from an oversampled band-limited signal. *IEEE Transactions on ASSP*, 31(3):752–755, 1983.

[15] R. Steele and F. Benjamin. Sample reduction and subsequent adaptive interpolation of speech signals. *Bell Systems Technical Journal*, 62(6):1365–1398, 1983.

[16] S.M. Kay. Some results in linear interpolation theory. *IEEE Transactions on ASSP*, 31(3):746–749, 1983.

[17] M. Pavon. Optimal interpolation for linear stochastic systems. *SIAM J. on Control and Optimization*, 22(4):618–629, 1984.

[18] M. Kohlmann and M. Pavon. Optimal interpolation for linear stochastic systems: the discrete time case. In *Proceedings of the 23rd IEEE Conference on Decision and Control*, pages 1479–1483, Las Vegas, 1984.

[19] G.B. Lockhart and D.J. Goodman. Reconstruction of missing speech packets by waveform substitution. In *Signal Processing 3: Theories and Applications*, pages 357–360, North-Holland, Amsterdam, 1986.

[20] H.-J. Platte and V. Rowedda. A burst error concealment method for digital audio tape application. *AES Preprint*, (2201):1–16, 1985.

[21] M.B. Priestley. *Spectral Analysis and Time Series*. Academic Press, London, 1981.

[22] G.H. Golub and C.F. van Loan. *Matrix Computations*. North Oxford Academic Publishing, Oxford, 1983.

[23] A. Papoulis. *Probability, Random Variables, and Stochastic Processes*. McGraw-Hill/Kogakusha, Tokyo, 1965.

[24] N. Wiener. *The Extrapolation, Interpolation and Smoothing of Stationary Time Series with Engineering Applications*. Wiley, New York, 1949.

[25] A. Kolmogorov. Interpolation and extrapolation. *Bull. Acad. Sci., USSR, Ser. Math.*, 5(4):3–14, 1941.

[26] C.L. Lawson and R.J. Hanson. *Solving Least Squares Problems*. Prentice Hall, Englewood Cliffs, New Jersey, 1974.

[27] M. Marcus and H. Minc. *Introduction to Linear Algebra*. MacMillan, New York, 1965.

[28] U. Grenander and G. Szegö. *Toeplitz Forms and their Applications*. University of California, Berkeley and Los Angeles, 1958.

[29] I.I. Hirschmann Jr. Recent developments in the theory of finite toeplitz operators. In *Advances in Probability and Related Topics*, Vol. 1, pages 105–167, Dekker, New York, 1971.

[30] S.L. Marple, Jr. *Digital Spectral Analysis with Applications*. Prentice Hall, Englewood Cliffs, New Jersey, 1987.

[31] J. Makhoul. Linear prediction: a tutorial review. *Proceedings of the IEEE*, 63(4):561–580, 1975.

[32] J.H. Wilkinson. Error analysis of direct methods of matrix inversion. *J. Assoc. Comp. Mach.*, 8:281–330, 1961.

[33] S.M. Kay and S.L. Marple, Jr. Spectrum analysis - a modern perspective. *Proceedings of the IEEE*, 69(11):1380–1419, 1981.

[34] L.R. Rabiner and R.W. Schafer. *Digital Processing of Speech Signals*. Prentice Hall, Englewood Cliffs, New Jersey, 1978.

[35] A.P. Dempster, N.M. Laird, and D.B. Rubin. Maximum likelihood from incomplete data via the EM algorithm. *Journal of the Royal Statistical Society, Series B*, 39:1–38, 1977.

[36] C.F.J. Wu. On the convergence properties of the EM algorithm. *The Annals of Statistics*, 11(1):95–103, 1983.

[37] R.A. Bayles. On the convergence of the EM algorithm. *Journal of the Royal Statistical Society, Series B*, 45(1):47–50, 1983.

[38] H. Akaike. A new look at the statistical model identification. *IEEE Transactions on Automatic Control*, 19(6):716–728, 1974.

[39] G. Dahlquist and Å. Björk. *Numerical Methods*. Prentice Hall, Englewood Cliffs, New Jersey, 1974.

[40] F. Itakura and S. Saito. A statistical method for estimation of speech spectral density and formant frequencies. *Electrical Communications in Japan*, 53-A(1):36–43, 1970.

[41] D. Seynhaeve and K. Swings. 'Architektuurontwerp van een digitale signaalverwerker voor adaptieve interpolatie van digitale audiosignalen', in Dutch. Master's thesis, Katholieke Universiteit Leuven, 1987.

[42] J.H. Wilkinson. *Rounding Errors in Algebraic Processes*. Prentice Hall, Englewood Cliffs, New Jersey, 1963.

[43] G. Cybenko. The numerical stability of the Levinson-Durbin algorithm for Toeplitz systems of equations. *SIAM J. on Sci. Stat. Comp.*, 1:303–320, 1980.

[44] J.B.H. Peek. *The Measurement of Correlation Functions in Correlators Using Shift-Invariant Independent Functions'*. PhD thesis, Technische Hogeschool Eindhoven, 1967.

[45] J.H. Wilkinson. *The Algebraic Eigenvalue Problem*. Clarendon Press, Oxford, 1965.

[46] N. Levinson. The Wiener RMS (Root Mean Square) error criterion in filter design and prediction. *J. Math. Phys.*, 25(4):261–278, 1947.

[47] P. Delsarte, Y. Genin, and Y. Kamp. A polynomial approach to the generalized Levinson algorithm, based on the Toeplitz distance. *IEEE Transactions on IT*, 29:268–278, 1983.

[48] A.W.M. van den Enden and N.A.M. Verhoeckx. *Discrete-Time Signal Processing: An Introduction*. Prentice Hall, Englewood Cliffs, New Jersey, 1989.

[49] L.R. Rabiner and B. Gold. *Theory and Application of Digital Signal Processing*. Prentice Hall, Englewood Cliffs, New Jersey, 1975.

[50] D. Slepian. Prolate spheriodal wave functions, Fourier analysis, and uncertainty - V: The discrete case. *Bell Systems Technical Journal*, 57:1371–1429, 1978.

[51] D.W. Tufts and R. Kumaresan. Estimation of frequencies of multiple sinusoids: making linear prediction perform like maximum likelihood. *Proceedings of the IEEE*, 70(9):975–989, 1982.

[52] H. Chen, T.K. Sarkar, S.A. Danant, and J.D. Brulé. Adaptive spectral estimation by the conjugate gradient method. *IEEE Transactions on ASSP*, 34(2):272–284, 1986.

[53] P.A. Maragos, R.W. Schafer, and R.M. Mersereau. Two-dimensional linear prediction and its application to adaptive predictive coding of images. *IEEE Transactions on ASSP*, 32(6):1213–1229, 1984.

[54] D.E. Dudgeon and R.M. Mersereau. *Multidimensional Signal Processing*. Prentice Hall, Englewood Cliffs, New Jersey, 1984.

[55] R.N.J. Veldhuis. Adaptive interpolation of autoregressive processes: a bound on the restoration error. *IEEE Transactions on ASSP*, 37(9):1462–1464, 1989.

[56] S.M. Kay. Recursive maximum likelihood estimation of autoregressive processes. *IEEE Transactions on ASSP*, 31(1):56–65, 1983.

[57] P.K. Rajan and H.C. Reddy. Formulation of 2-D normal equations using 2-D to 1-D form preserving transformations. *IEEE Transactions on ASSP*, 36(3):414–417, 1988.

Index